Silurian nuculoid and modiomorp from Sweden

LOUIS LILJEDAHL

Liljedahl, L. 1994 04 15: Silurian nuculoid and modiomorphid bivalves from Sweden. *Fossils and Strata*, No. 33, pp. 1–89. Oslo. ISSN 0300-9491. ISBN 82-00-37648-6.

The nuculoids and modiomorphids from the Silurian of Sweden are reviewed with respect to their taxonomy, functional morphology, ecology, and stratigraphical distribution. Several thousands of specimens, in many cases excellently preserved, have been studied. Twenty-one species of nuculoids, of which *Praenucula faba* and *Ekstadia kellyi* are new, assigned to 10 genera, are referred to the families Praenuculidae and Nuculidae of the Nuculacea; Malletiidae and Nuculanidae of the Nuculanacea. The extensive nuculoid material available enabled ontogeny and intraspecific variation to be recorded for some of the taxa. Three nuculoid associations have been studied, in which tiering and life habits of the different species have been reconstructed. The nuculoid species have a stratigraphical range from the uppermost Llandovery to uppermost Ludlow. Some 25 modiomorphid species have been recognized, of which 18 are new, belonging to 7 genera, of which *Aleodonta* and *Mimerodonta* are new. The family Modiomorphidae is subdivided into the new subfamilies Modiomorphinae and Modiolopsinae. The species recorded exhibit a muscular morphology pattern intermediate between those of highly effective burrowers and Recent myti-lids. Most species also show indications of byssal attachment. Thus, it appears as if the Silurian modiomorphids examined retained the ability of effective burrowing. Several of them also had a byssus for maintaining a stable life position. These bivalves have been found in various types of sediments, from fine-grained limestones and siltstones to coarse-grained oolitic limestones and sandstones, indicating a variety of habitats. The species have a stratigraphical range from the uppermost Llandovery to Pridoli. Three of the seven genera seem to have been endemic to Baltoscandia in Silurian times. □*Bivalvia, Nuculoida, Mytiloida, Modimorphidae,* Aleodonta, *taxonomy, functional morphology, life habits, ecology, Silurian, Gotland, Scania, Sweden.*

Louis Liljedahl, Department of Historical Geology and Palaeontology, Sölvegatan 13, S-223 62 Lund, Sweden; received 10 June, 1991; revised manuscript received 18 September, 1992.

Contents

Introduction

Despite their being characteristic elements in many faunas, Silurian bivalves are seldom mentioned in the geological or palaeontological literature. This is due to the fact that they have to date received insufficient study.

The present paper is based on a restudy of all hitherto described nuculoids and modiomorphids from the Silurian of Sweden as well as on old and new collections of thousands of specimens. Also, pertinent material from institutions in Europe, North America and Australia has been studied.

Living nuculoids exhibit much the same shell morphologies as their fossil counterparts. Thus, the functional morphology of Palaeozoic nuculoids is fairly well understood. However, the often simple and uniform morphological appearance makes classification arduous, especially when based on external features alone. This is often the case with Palaeozoic nuculoids, as the fossil specimens are usually preserved as articulated shells.

The Silurian nuculoids of Gotland are generally found in marls and are comparatively well preserved. A few localities have yielded silicified specimens, representing by far the best preserved Palaeozoic bivalve faunas ever recorded.

Most descriptions of modiomorphids are from the Ordovician of North America and Australia and the Devonian of North America, and only a few scattered reports deal with these bivalves from the Silurian of Europe, North America, and Australia.

The modiomorphids discussed herein form morphologically and ecologically the most diverse group of bivalves from the Silurian of Sweden. They were suspension feeders, exhibit a variety of life habits and adaptations to a wide range of environments, as is indicated by their shell morphology in combination with sedimentological and other geological data. Thus, they provide relatively simple and reliable palaeoecological tools that should prove to be of value to other students working on palaeontology and geology of Silurian rocks.

Geology

The Silurian strata of Gotland consist of a series of mainly limestones and calcareous shales ranging in age from latest Llandovery in the northwestern part to latest Ludlow in the south. The sequence is approximately 500 m thick and was deposited when the area was close to the Silurian equator (Creer 1973), and at a water depth not exceeding 175–200 m (Gray, Laufeld & Boucot 1974).

The Silurian sequence of Gotland was subdivided by Hede (1921, 1925) into 13 units (Fig. 1) which have been used and refined by various authors (e.g., Hede 1960; Martinsson 1962, 1967; Manten 1971; Laufeld 1974; Jeppsson 1974, 1983; Larsson 1979).

A presentation of the Silurian of Scania is shown in Fig. 1. For detailed discussions, see, e.g., Martinsson (1962) and Laufeld & Jeppsson (1976).

Material, state of preservation, and preparation of fossils

The bulk of the present material belongs to the Swedish Museum of Natural History, Stockholm and the remaining part to the Geological Survey of Sweden, Uppsala, the Palaeontological Institution, Uppsala, and the Department of Historical Geology and Palaeontology, Lund.

Much of the Swedish Museum of Natural History's material was collected by G. Lindström in the middle of the 19th century. He was especially interested in the bivalves from Gotland and planned to monograph these fossils. This intention, however, was never fulfilled.

After having finished his work on the Ordovician bivalves from the Leptaenakalk in Dalarna (now Boda Limestone), O. Isberg began to study the fossil bivalve material of Gotland collected mainly by Lindström and J.E. Hede. Isberg sorted

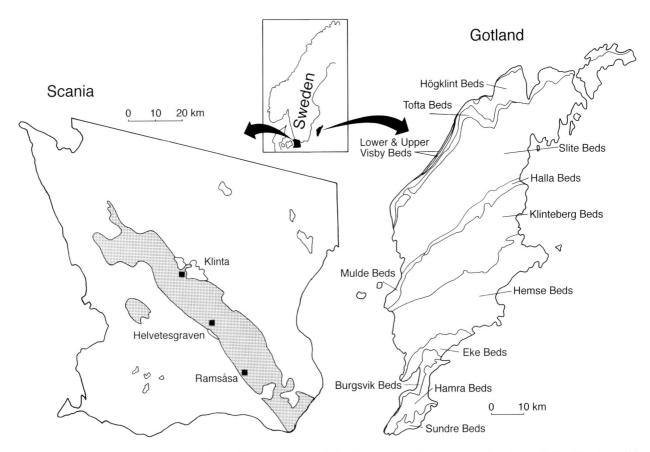

Fig. 1. Simplified geological map of Scania showing the Silurian outcrop (stippled) and sampled localities (black squares), and map of Gotland showing Hede's (1921 and 1925) units (cf. Figs. 26–27).

and prepared many specimens and even made some plates, but this work was not completed. No written information accompanying these plates has been found.

The specimens in the old collections of the institutions just mentioned consist of recrystallized calcium carbonate, and represent internal and external impressions, composite moulds, as well as steinkerns. Part of the study material is silicified, and it has been extracted by means of acid etching as part of a project on silicified fossils from Gotland. Samples, in some cases with a weight of 50 kg, normally about 5–20 kg, were etched in 6% acetic acid and 2% calcium acetate (without the calcium acetate the solution is corrosive to phosphatic fossils; see Jeppsson *et al.* 1985).

When the calcium carbonate was completely dissolved, the often delicate silicified fossil remains were picked, washed, and dried. Extremely thin and fragile specimens were treated in a solution of polyvinylacetate and acetone. It must be noted that a too thick solution effaces the surface topography of the fossils, and it must temporarily be removed by acetone prior to photography. For the handling of silicified fossils, a pair of entomologist's tweezers is neccessary.

Prior to photography the specimens were blackened with fountain-pen India ink and coated with ammonium chloride.

Repositories

Specimens with numbers prefixed 'RMMo' are deposited in the collections of the Swedish Museum of Natural History, Box 50007, S-104 05 Stockholm, Sweden, those prefixed 'SGU Type' are deposited in the type collection of the Geological Survey of Sweden, Box 670, S-751 28 Uppsala, Sweden, and those prefixed 'LO' in the type collection of the Geological Institute, Lund University, Sölvegatan 13, S-223 62 Lund, Sweden.

Orientation, measurements, and terminology

Nuculoida

The following criteria may be used as means for orientation of bivalves (Cox 1969; Bradshaw & Bradshaw 1971, 1978; Driscoll 1964): If ligamental nymphs are present, they are invariably posterior to the beaks. If a pallial sinus is present, it is invariably near the posterior end. A rostrate end is usually the posterior one. A single diagonal sulcus is usually situated

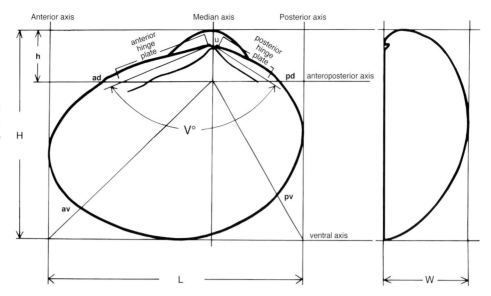

Fig. 2. Terminology and measurements of nuculoid shell. The part between *ad* and *pd* is the dorsal margin, *u* to *ad* anterior part of dorsal margin, *ad* to *av* anterior margin, *av* to *pv* ventral margin, *pv* to *pd* posterior margin, and *pd* to *u* posterior part of dorsal margin; V = hinge angle, H = total height, h = hinge height, L = total length, and W = width.

in the posterior part of the valve. If there is a considerable difference in morphology and size between the teeth of the two hinge plates, the anterior ones are arcuate and larger and higher than the posterior ones. The attachment area for the gonads is in the anterior part of the shell. An oblique ridge close to the posterior adductor muscle scar may reflect the position of gill attachment muscles. The configuration of the pedal and accessory muscles may indicate the anterior end where the foot protrudes.

Following the example of Liljedahl (1983), the antero-posterior axis (ap-axis) is defined as the line, p, going through the distalmost tooth in each hinge plate. In cases where the hinge is not observable, the ap-axis coincides with the line of maximum extension of the shell. Perpendicular to the ap-axis are the anterior axis (tangential to the anterior edge of the shell), the posterior axis (tangential to the posterior edge of the shell), and the median axis. The ventral axis is parallel to the ap-axis and tangential to the shell at the extreme point of the ventral margin. The total length of the shell (L) is measured from the anterior axis to the posterior axis, and the total height (H) from the highest point of umbo to the ventral axis. The width (W) of each valve is measured from the sagittal plane to the point of maximum convexity, laterally. The hinge angle (V) is measured from the tip of the beak to the distalmost tooth of each hinge plate (Fig. 2).

Modiomorphidae

The anterior end of modiomorphids is hypertrophied and easily recognizable. Generally accepted bivalve orientation for measuring based on soft parts and their scars (Fischer 1886, antero-posterior axis touching lower margin of the adductors) or on ligament extensions (e.g., Baily 1983), is not used here because most specimens measured consist of articulated valves or internal moulds with internal ligament.

Instead, orientation of the shell is based mainly on external features. The hinge line of modiomorphids is convex and is therefore unsuitable as a line of orientation for measuring. Measurements made herein are illustrated in Fig. 3. Thus, total length of the shell (L) is equal to greatest extension of the shell and total height (H) corresponds to the maximum extension perpendicular to the length direction. The length-axis thus obtained does not coincide with the antero-posterior axis, which is generally accepted as touching the lower margin of the two adductor muscle scars. The width (W) is measured from the point of maximum convexity of each valve perpendicular to the length in a horisontal plane.

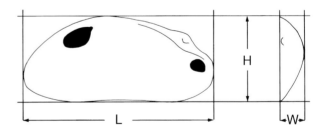

Fig. 3. Measurements of the modiomorphid shell. L = maximum extension, H = height, W = width (for definitions see text).

Morphology and biology

Nuculoida

Morphology

General shell shape. – The shell of the Gotland nuculoids may be either anteriorly expanded (e.g., *Nuculoidea*, Fig. 30C), posteriorly expanded (e.g., '*Nuculana*' *oolitica*, Fig. 35I, K), or equilateral (*Similodonta*, Fig. 29I). Shell shape in lateral outline varies from almost circular (*Similodonta*) to elliptical elongate ('*Nuculana*'), and the valves may be compressed (e.g., *Similodonta*, Fig. 29) or gibbous (e.g., *Nuculodonta*, Fig. 28D, K).

The umbo may be pronounced (e.g., ?*Nuculoidea* sp., Fig. 31Q) or low and inconspicuous (e.g., ?*Nuculoidea ecaudata*, Fig. 32), and may bear more or less defined ridges, one (e.g., *Tancrediopsis*, Fig. 35M) or three (e.g., *Ekstadia*, Fig. 34B) in number, and may lack a defined umbonal ridge (e.g., *Similodonta*).

Some shells may exhibit a diagonal siphon-indicative sulcus, which extends from the umbo to the posterior part of the ventral margin (e.g., *Ekstadia tricarinata*, Fig. 34A, B, and *Palaeostraba baltica*, Fig. 34O).

Shell sculpture. – In some species the shell surface consists of faint growth lines on a smooth surface (e.g., *Nuculoidea lens*, Fig. 30C), sometimes with evident growth-increment stops (e.g., *Similodonta*, Fig. 29I), rarely with more conspicuous stops reflecting longer periods of retarded growth (e.g., *Nuculodonta gotlandica*, Fig. 28H). In some species (e.g., *Nuculoidea* sp. A, Fig. 31O) the sculpture may also consist of evident concentric ribs or of coarse lamelliform ridges (*Tancrediopsis solituda*, Fig. 35M).

Size. – The present material includes the smallest Palaeozoic nuculoids hitherto recorded; they are represented by juvenile specimens, measuring about 0.5 mm in length (Liljedahl 1984a, Fig. 1E). The length of adult specimens of the different species varies from about 7 mm ('*Nuculana*' sp. A, Fig. 35I) to about 30 mm (*Caesariella lindensis*, Fig. 33B).

External ligament. – Some of the nuculoid taxa have ligamental nymphs, forming the site of an external, opisthodetic ligament (e.g., *Caesariella lindensis*, Fig. 33J). In some cases, however, fossilized remnants of the ligament tissue are present (e.g., *Tancrediopsis gotlandica*, Fig. 35L).

Hinge. – The nuculoid hinge consists of two series of teeth, posterior and anterior to the beak, respectively.

In shells with an external ligament the two hinge plates bear teeth that are different in size and shape, the anterior teeth being large, often chevron-shaped, and the posterior ones lower, narrower and of varying shape (e.g., *Praenucula faba*, Fig. 29A). In some of these species, the two series of hinge teeth are interrupted by a resilifer-like structure (e.g., *Nuculodonta gotlandica*, Fig. 28J), while in others the tooth rows continue into each other (e.g., *Caesariella lindensis*, Fig. 33J.

In shells with a resilifer, indicating the presence of an internal ligament, the teeth of the two hinge plates are often more uniform (e.g., '*Nuculana*' *oolitica*, Fig. 35J).

Ligament. – A resilifer is indicative of an internal ligament (e.g., *Nuculoidea lens*, Fig. 30F, G, J). Rarely remnants of the ligament tissue proper are preserved (see, however, silicified replica of ligament of *Nuculoidea lens* in Liljedahl 1983, p. 20, Fig. 5D).

Some shells of *Nuculodonta gotlandica* (which had an external ligament), also exhibit a resilifer-like structure with a non-denticulate area above separating the two rows of hinge teeth (Fig. 28J).

In addition, *Caesariella lindensis* (Fig. 33J) and *Palaeostraba baltica* (Fig. 34P) have a non-denticulate area just below the region where the two series of teeth meet, possibly indicating an internal part of the ligament.

Hinge reinforcements. – Some shells possess hinge-plate reinforcements. *Palaostraba baltica* has two slender ridges, the one anteriorly and the other posteriorly (Fig. 34L), while *Nuculites solida*, *Nuculites* sp. A and *Nuculites* sp. B exhibit a robust septum in the anterior of the shell (Fig. 35A, D, G).

Musculature. – The adductor muscle scars are subequal in the investigated nuculoids. The anterior adductor muscle is invariably somewhat larger and more impressed than the posterior one (Fig. 30A, 31J). The posterior pedal-retractor muscle scar is faint and rarely distinguishable (Fig. 30D, G), whereas the anterior pedal-protractor muscle scar and the anterior pedal-retractor–elevator muscle scar are often conspicuously incised (Fig. 28I, 30F). Visceral attachment muscle scars are rarely preserved but may be evident (Fig. 30F, 31J, 34F). Marks indicative of the visceral floor are also rare (Figs. 33F, 34L).

Bradshaw (1978, p. 213) argued that Ordovician nuculoids lack evidence for the existence of a muscular visceral floor. By the Devonian, however, the pattern of umbonal muscles changed significantly, and in species of *Nuculoidea* traces of a possible, primitive, umbonal visceral floor were identified (Bradshaw 1978). If the interpretation of the umbonal features in *Caesariella* and *Palaeostraba* of the present material is correct, a visceral floor was developed already in Silurian times.

Sinuation of the pallial line, indicating the presence of siphons, has been observed i, e.g., *Ekstadia tricarinata* (Fig. 34E) and *Ekstadia kellyi* (Fig. 34I).

Ontogeny

The prodissoconch II (Cox 1969) of Recent bivalves is generally 0.2–0.6 mm long and usually has insufficient characters for systematic determination (LePennec 1978, pp. 31, 211). In a few cases, however, there are differences in morphology of the prodissoconch II between separate taxa, e.g., *Nucula* and *Yoldia* (Drew 1901, p. 340). When the shell has grown to 2–3 mm in length, the morphology of the hinge teeth may indicate which species it belongs to (LePennec 1978, p. 213).

The rich, silicified nuculoid fauna of Möllbos 1 on Gotland has provided specimens ranging from 0.5 mm to 15.0 mm in length (Liljedahl 1983, 1984a). Specimens shorter than 2.0 mm generally lack diagnostic characters, whereas shells of *Nuculodonta gotlandica* slightly exceeding this size may be identified by their conspicuous external ligament.

The mean height/length ratio of specimens of *Nuculoidea lens* below 3.0 mm in length is 0.87, and of *Nuculodonta gotlandica* 0.82. With increasing shell size, the shell shape of both species changes in lateral outline from almost circular to ovate, which is most accentuated in *Nuculodonta gotlandica* (Liljedahl 1983, p. 21, Fig. 6). The smallest identifiable shell of *Nuculodonta gotlandica* is 1.5 mm long and of *Nuculoidea lens* 1.8 mm.

The number of hinge teeth increases with growth. The smallest shell with countable teeth of *Nuculoidea lens*, ca. 3.0 mm long, has an equal number of teeth on each hinge plate, approximately 5–6 teeth. The largest specimen of this species, ca. 15.0 mm long, contains approximately 15 teeth in the anterior hinge plate and ca. 11 in the posterior one (Liljedahl 1983, p, 23, Fig. 7). The smallest specimen of *Nuculodonta gotlandica* with countable hinge teeth, approximately 5.0 mm long, contains in the anterior hinge plate ca. 14 teeth and in the posterior ca. 8 teeth; the largest specimens, approximately 15.0 mm, ca. 24 teeth in the anterior and ca. 14 teeth in the posterior hinge plate (Liljedahl 1983, p. 36, Fig. 12).

Functional morphology

In shells of living bivalves, traces of soft parts are generally faint or completely lacking. Palaeozoic species, however, commonly show evident muscular impressions, displaying elaborate patterns of pedal, visceral, and branchial muscles.

As far as the nuculoids are concerned, the commonly notable lack of muscle scars in Recent species is due to the large attachment area of the muscles, which thus compensates for a shallower relief in the shell (cf. Driscoll 1964, p. 65).

Although there seems to have been some evolutionary changes in the musculature set-up in nuculoids (see, e.g., Bradshaw 1978), many Palaeozoic nuculoids show a general pattern of muscle scars similar to the pedal muscles and accessory muscles in species of living nuculoid genera (Figs. 4, 6, 7, and 8 herein; see also, e.g., Driscoll 1964; Bradshaw 1974, 1978; Liljedahl 1983, 1984a)

The movements of the nuculoid foot are performed by interacting protraction and retraction muscles, respectively, in a simple mechanical way (Fig. 5). Hence, using the positions of these scars as a guide, it seems to be possible to infer the direction of foot protrusion (Driscoll 1964), and also the size of the foot and resulting burrowing ability of the species.

Muscular impressions have been recorded in 8 of the 21 species described here. The majority of the shells examined show more or less evident impressions of the adductor muscles. In a few shells scars of other muscles and of additional soft parts have been recorded.

Nuculoidea lens (Fig. 30F, G, J) and *Nuculoidea pinguis* (Fig. 31B, CJ) show an almost identical pattern of muscular impressions, viz. with a relatively large anterior protractor muscle scar and large median muscle scars. The latter most probably indicate the site of muscles going down into the foot and also in the visceral floor, reinforcing both the pedal retractors and also anchoring the pericardial region to the

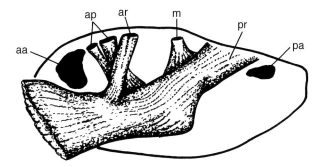

Fig. 4. Main pedal muscles and adductor muscles of the extant nuculoid *Yoldia limatula*. aa = anterior adductor, ap = anterior pedal protractors, ar = anterior pedal retractor, m = median muscle, pr = posterior pedal retractor, pa = posterior adductor (after Heath 1937, Pl. 10:83).

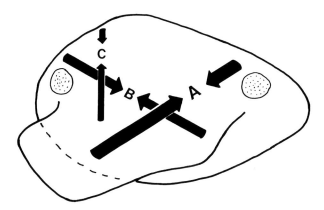

Fig. 5. Semidiagrammatic representation of the simple mechanics of the pedal muscles of the foot in nuculoids (anterior to the left). Contraction along 'A' results in tension and retraction of the foot. Contraction along 'B' results in protrusion of the foot anteriorly. Contraction along 'C' results in tension and retraction/elevation of the foot dorsally (after Driscoll 1964).

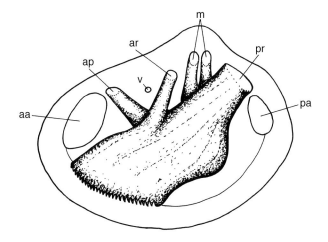

Fig. 6. *Nuculoidea lens*. Reconstruction of the foot and its muscles. Same lettering as in Fig. 4. v = visceral attachment muscle.

shell (cf. Heath 1937, p. 14). The scar, which by Liljedahl (1983, Fig. 8, No. 8 and 1984a, Fig. 3A, No. 8) was considered to be a pedal muscle scar (Fig. 6v), is now suggested to be an attachment scar for the visceral floor (cf. Bradshaw 1978;

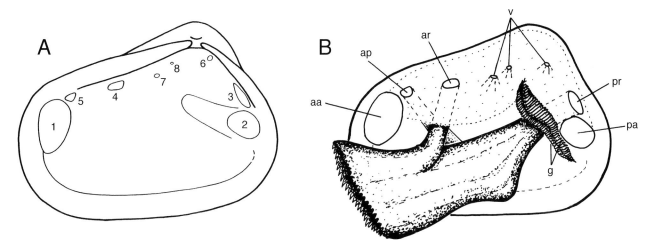

Fig. 7. Nuculodonta gotlandica Liljedahl, 1983. □A: Muscular imprints, anterior to the left. 1 = anterior adductor muscle scar, 2 = posterior adductor muscle scar, 3 = posterior pedal retractor muscle scar, 4 = anterior pedal retractor muscle scar, 5 = anterior protractor muscle scar, 6–8 = visceral attachment muscle scars (after Liljedahl 1983, Fig. 15). □B. Reconstruction of the foot and its muscles. Lettering as in Fig. 4. v = visceral attachment muscle. Stippled area indicating pericardial region.

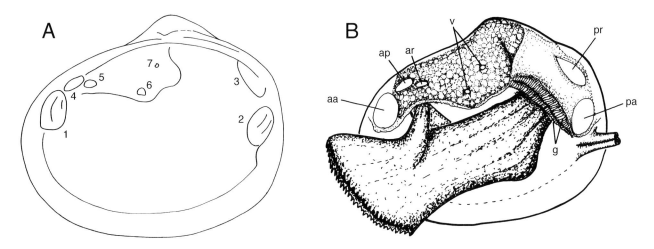

Fig. 8. Caesariella lindensis Soot-Ryen, 1964. □A: Muscular impressions (anterior to the left) 1 = anterior adductor muscle scar, 2 = posterior adductor muscle scar, 3 = posterior pedal retractor muscle scar, 4 = anterior pedal protractor muscle scar, 5 = anterior pedal retractor muscle scar, 6–7 = attachment muscle scars of the visceral floor. □B: Reconstruction of the foot and its muscles and extension of the gonads (anterior part of pericardial region) and locations of gills (g). Left view. Lettering same as in Fig. 4. v = visceral attachment muscles.

Heath 1937). In *Nuculoidea pinguis* two visceral attachment muscle scars have been observed. However, no anterior pedal retractor muscle scar is at hand in the present material of this species. The muscular system of *Nuculodonta gotlandica* (Fig. 7) deviates somewhat from that of the two first mentioned species. The scars nos. 6,7, and 8 of Liljedahl (1983, Fig. 15), then considered as median muscle scars, I suggest to be attachment scars for the visceral floor. This species also has a relatively large anterior retractor muscle scar and an oblique, smooth-edged ridge ventral and anterior to the posterior adductor muscle scar (Fig. 28N). The location of this ridge agrees with the position of the gills in Recent *Nucula* and *Acila* (Fig. 9) and possibly indicates the site of gill attachment muscles [cf. similar structure of the Triassic *Palaeonucula variabilis* (Sowerby) *in* Bradshaw 1978, Fig. 5B].

Caesariella lindensis (Fig. 8) has an unusually large anterior pedal protractor muscle scar, a conspicuous anterior pedal retractor muscle scar, and a similarly large pedal retractor muscle scar and evident attachment muscle scars of the visceral floor. This species also shows a conspicuous restricted area enclosing the anterior upper part of the umbonal cavity (Fig. 33F), probably showing the extension of the visceral floor (cf. similar features in Recent nuculoids, Fig. 9 herein; Heath 1937; Bradshaw 1978). The pallial line of *Caesariella lindensis* has a shallow sinus indicating the presence of short siphons.

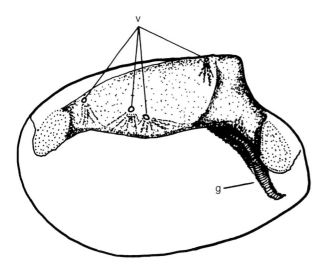

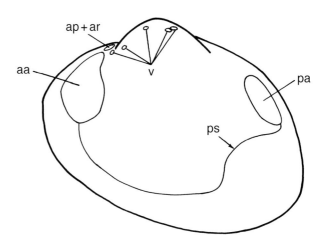

Fig. 9. Acila divaricata. Left valve removed showing pericardial region with gills (g) and attachment muscles (v) for the visceral floor (after Heath 1934, Pl. 4:37). Anterior to the left.

Fig. 10. Ekstadia kellyi n.sp. Muscular impressions (anterior to the left). Same lettering as in Fig. 4. v = visceral attachment muscles. ps = pallial sinus.

Ekstadia kellyi (Fig. 10) contains an elongated anterior pedal protractor muscle scar, and five visceral muscle scars. It has, like *Ekstadia tricarinata*, a relatively deep pallial sinus, suggesting comparatively long siphons.

Palaeostraba baltica has a faint impression in the centre of the shell, possibly being the site of a visceral attachment muscle. In the umbonal cavity there is a restricted depression (Fig. 34L), showing the possible extension of the visceral floor. The internal ridges on each side of the hinge (Fig. 34L) may have functioned as attachment area for visceral muscles. Two incisements in one specimen, which are situated between the posterior adductor muscle scar and the shell margin, are enigmatic and perhaps not muscular impressions at all (Liljedahl 1984a, Fig. 6).

Nuculites sp. A (Fig. 35D) has a subcircular anterior adductor muscle scar, which is limited posteriorly by the internal septum. It also contains two small punctiform scars in the umbonal cavity in an anterior position.

Life habits

The nuculoids are mainly deposit feeders, and most species therefore actively move about in the sediment which they devour. Most deposit-feeding bivalves are rapid burrowers, although many of these species are moderately rapid or slow burrowers (Stanley 1970). Besides an adequate shell shape, with a generally smooth surface facilitating burrowing, the exceptionally large and powerful foot of nuculoids is a prerequisite for effective locomotion.

General shell shape, shell thickness, surface ornamentation, etc., may give clues to the life position and feeding depth in the sediment (Stanley 1970). In several of the present species the anterior part of the shell is expanded, and together with the conspicuous scars of pedal muscles preserved in some species, this indicates a strong and powerful foot well suited for burrowing (Stanley 1970, p. 66). The anterior adductor muscle scar is, where observed, deep and invariably larger than the posterior adductor muscle scar, suggesting a powerful closure and thus efficient posterior removal of debris and other indigestible matters from the mantle cavity. It seems plausible that much of the mantle cavity was occupied by the foot and its muscles in these species, while the gills were most probably moderate in size. (See suggested reconstruction of the mentioned features of *Nuculodonta gotlandica* in Fig. 7B and of *Caesariella lindensis* in Fig. 8B)

A compressed and elongate shell form, like that of *Tancrediopsis gotlandica* and *Tancrediopsis solituda,* is normally indicative of rapid burrowing ability (and thus a deep life position). However, the thick shell with conspicuous concentric ornamentation in *Tancrediopsis gotlandica* and coarse sculpture in *Tancrediopsis solida* instead suggest a life position close to the sediment surface. The diagonal posterior umbonal keel in these species may even show the probable level at which the shell protruded from the sediment (Fig. 13:1,7).

Length, height and width of Recent bivalves were compared by Stanley (1970, Fig. 25) to show the relation of burrowing rate to gross shell shape. The values of all nuculoids examined are within the region of rapid burrowing, which is not surprising, since nuculoids normally are active bivalves moving about constantly.

The value of *Nuculodonta gotlandica* is fairly close to the region of slow burrowing (Fig. 11:1). The shell of this species is thick in comparison with that of the other species in the Möllbos bivalve association. In Stanley's review, virtually all thick-shelled species are shallow burrowers (Stanley 1970, p. 68). The occurrence of *N. gotlandica* in relation to other

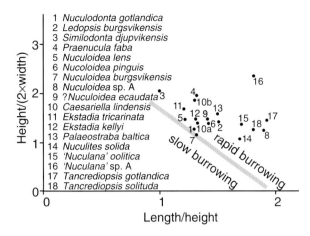

Fig. 11. Shell shape related to burrowing rate (after Stanley 1970, Fig. 25). (10 *Caesariella lindensis,* a = Ludlovian variety, b = Wenlockian variety; see discussion under 'Life habits')

bivalves from Möllbos suggests, together with the morphological features mentioned, that it had a life position just beneath the sediment surface (Fig. 12:1).

Ledopsis burgsvikensis also has a rather thick shell and a triangular outline. It is placed close to the boundary between slow burrowing and rapid burrowing (Fig. 11:2). I suggest that it lived close to the sediment surface (Fig. 14:1)

Similodonta djupvikensis has a compressed, thin shell with a smooth surface and an almost circular outline. The values of gross shell shape lie fairly close to the boundary between slow

and rapid burrowing (Fig. 11:3). The conspicuously enlarged anterior part of the shell of this species suggests a large and effective foot for burrowing and speaks in favour of a relatively deep life position (Fig. 13:2).

The shell of *Praenucula faba* is compressed, anteriorly expanded and the shell surface is smooth. Shell shape indicates a moderately rapid burrowing ability (Fig. 11:4) and a large foot (conspicuously extended anterior end). I suggest that it had a life position relatively deep in the sediment (Fig. 13:8).

In Fig. 11, the values of length, height, and width of *Nuculoidea lens* (Fig. 11:5) fall close to the region of slow burrowing. The shell is relatively thick and it has a smooth surface. There are conspicuous impressions of pedal muscles suggesting a powerful foot. The morphology thus indicates, together with statistical facts (see below), an active mode of life in a relatively shallow life position (Fig. 12:2).

Nuculoidea pinguis is closely similar in shell morphology and muscular impressions to *Nuculoidea lens* and, accordingly, most probably had the same or very similar burrowing ability (Fig. 11:6), life habit and life position as this species (Fig. 13:5).

Although no muscular impressions have been observed in *Nuculoidea burgsvikensis,* the close similarity in shell morphology to *Nuculoidea lens* and *Nuculoidea pinguis* suggests a similar life habit and burrowing ability (Fig. 11:7). Possibly the life position of *Nuculoidea burgsvikensis* was somewhat shallower than that of the other two species mentioned, owing to its comparatively thicker and coarser shell (Fig. 14:4).

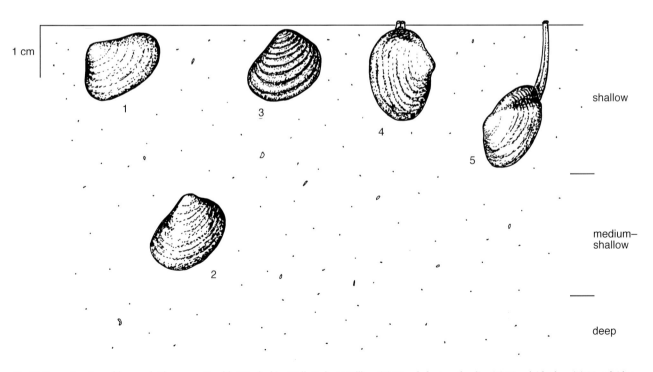

Fig. 12. Reconstruction of the nuculoid community of the Wenlockian Halla Beds at Möllbos. (1) *Nuculodonta gotlandica,* (2) *Nuculoidea lens,* (3) *Nuculoidea* sp. A, (4) *Caesariella lindensis,* (5) *Palaeostraba baltica.*

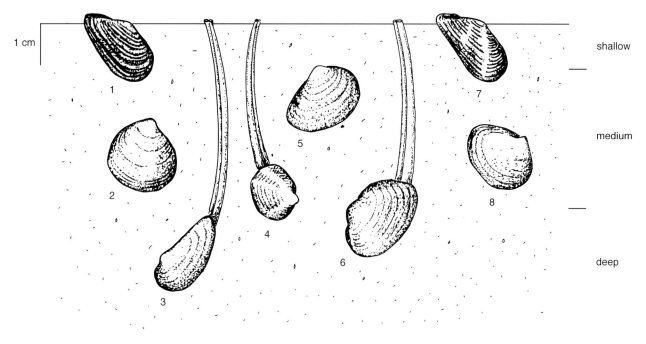

Fig. 13. Reconstruction of the nuculoid community of the Wenlockian Mulde Beds at Djupvik. (1) *Tancrediopsis solituda*, (2) *Similodonta djupvikensis*, (3) '*Nuculana*' sp. A, (4) *Ekstadia tricarinata*, (5) *Nuculoidea pinguis*, (6) *Ekstadia kellyi*, (7) *Tancrediopsis gotlandica*, (8) *Praenucula faba*.

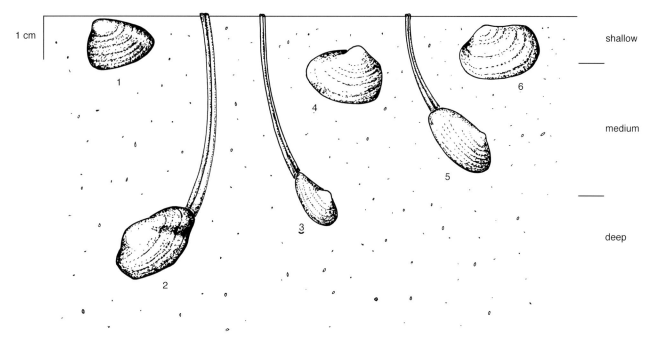

Fig. 14. Suggested life positions of the nuculoid species of the Ludlovian Burgsvik Beds. (1) *Ledopsis burgsvikensis*, (2) *Nuculites solida*, (3) '*Nuculana*' oolitica, (4) *Nuculoidea burgsvikensis*, (5) *Nuculites* sp. B, (6) ?*Nuculoidea ecaudata*.

The shape of the shell in *Nuculoidea* sp. A points to a moderately rapid burrowing speed (Fig. 11:8). Furthermore, it has a conspicuously coarse shell sculpture and a thick and gibbous shell, both characteristics of a shallow life position (Fig. 12:3).

?*Nuculoidea ecaudata* has an anterior expansion of the shell and a rather thick shell, and its values of length vs. height and width indicate a moderately rapid burrowing rate (Fig. 11:9). The thick shell is indicative of a shallow life position (Fig. 14:6).

Two varieties of *Caesariella lindensis* can be recognized, viz. the Ludlovian type from the Slite Beds and the Wenlockian type from the Halla Beds. The first-mentioned lies close to the region of slow burrowing and the other well withing the region of rapid burrowing (Fig. 11:10a and 11:10b, respectively).

The value of length, height, and width of *Ekstadia tricarinata* indicates moderately rapid burrowing (Fig. 11:11), which is also true of *Ekstadia kellyi* (Fig. 11:12). The corresponding values of *Palaeostraba baltica* and *Nuculites solida*, respectively, lie well within the region of rapid burrowing (Fig. 11:13 and 11:14, respectively), indicating a somewhat faster burrowing speed than the two *Ekstadia* species. The values of *'Nuculana' oolitica* (Fig. 11:15), *Tancrediopsis gotlandica* (Fig. 11:17) and *Tancrediopsis solituda* (Fig. 11:18) are well within the region of rapid burrowing. The value of *'Nuculana'* sp. A lies deepest in the region of rapid burrowing (Fig. 11:16).

Ekstadia tricarinata, Ekstadia kellyi, Palaeostraba baltica, Nuculites solida, Nuculites sp. A, *Nuculites* sp. B, *Caesariella lindensis, 'Nuculana' oolitica,* and *'Nuculana'* sp. A are all considered to have had siphons. In *Ekstadia tricarinata, Ekstadia kellyi, Nuculites* sp. A, and *Caesariella lindensis* the presence of a pallial sinus supports this assumption. *Palaeostraba baltica* and *Nuculites solida* both have an oblique posterior sulcus, indicative of siphons, whereas in *'Nuculana' oolitica* and *'Nuculana'* sp. A the conspicuously rostrate posterior end suggests the presence of siphons. I suggest that the siphonate species, except for *Caesariella lindensis* (see discussion below), had a comparatively deep life position (*Palaeostraba baltica*, Fig. 12:5; *'Nuculana'* sp. A, Fig. 13:3; *Ekstadia tricarinata*, Fig. 13:4; *Ekstadia kellyi*, Fig. 13:6; *Nuculites solida*, Fig. 14:2; *'Nuculana' oolitica*, Fig. 14:3; *Nuculites* sp. B, Fig. 14:5).

All of the present nuculoids fall within the region of rapid burrowing in Fig. 11. One species, however, *'Nuculana'* sp. A (Fig. 11:16), exhibits an extreme, blade-like shell shape, indicating extraordinarily rapid burrowing. The remaining species were probabaly rapid to moderately rapid burrowers, and their positions as shown in Fig. 11 may give a rough idea of their burrowing ability in relation to each other, especially of those occurring in the same habitat (see below).

Ecology and faunal associations

Living nuculoids are found as infauna in most sublittoral environments, from gravelly to muddy bottoms, in which each species is generally restricted to a particular type of substratum (Allen 1954, p. 469). The Silurian nuculoids of Gotland have been recorded from a wide range of substrates, from coarse oolitic sandstone to argillaceous limestones and muds. Although disarticulated, the nuculoids found in coarse sediments, such as oolitic limestones and sandstones, probably lived in these habitats, because the shells have not been eroded.

		Non-siphonate		Siphonate
Fine-grained	Shallow	*Tancrediopsis gotlandica* *Tancrediopsis solituda* *Nuculodonta gotlandica* *Nuculoidea* sp. A	Shallow	*Caesariella lindensis* *Palaeostraba baltica*
	Medium	*Nuculoidea lens* *Nuculoidea pinguis* *Similodonta djupvikensis* *Praenucula faba*	Medium	*Ekstadia tricarinata* *Ekstadia kellyi*
	Deep		Deep	*'Nuculana'* sp. A *Nuculites* sp. A
Coarse-grained	Shallow	*Ledopsis burgsvikensis* *Nuculoidea burgsvikensis* ?*Nuculoidea ecaudata*	Shallow	
	Med.		Med.	*Nuculites* sp. B
	Deep		Deep	*'Nuculana' oolitica* *Nuculites solida*

Fig. 15. Mode of feeding (non-siphonate = deposit feeding, siphonate = suspension feeding), maximum feeding depth and substrate preference of the nuculoids of Gotland.

Based on morphological features, 11 of the present species are considered non-siphonate and 9 siphonate (Fig. 15). They are grouped in relation to grain-size of the substrate and to maximum feeding depth.

So far, three nuculoid associations from the Silurian of Gotland have been recorded, viz. the Möllbos fauna (Halla Beds), the Djupvik fauna (Mulde Beds), and the Burgsvik fauna (Burgsvik Beds). The bivalve fauna from Möllbos 1 is exclusively silicified (Liljedahl 1983, 1984a, b, c, 1985, 1986, 1989a, b), whereas the remaining faunas are preserved as calcitic shells or moulds of varying material.

The Möllbos fauna. – At Möllbos 1 (Fig. 12), 90% of the individuals are deposit-feeders, mainly nuculoids (Liljedahl 1985, pp. 61–64, Figs. 6–7). The semi-infaunal suspension-feeders constitute 10% and epifaunal suspension-feeders only 0.03% of the individuals. The extremely muddy habitat of Möllbos appears to have favoured deposit-feeders, whose activities disturbed the feeding of the suspension-feeders. Also, non-siphonate deposit-feeders probably disturbed the feeding of the siphonate deposit-feeders (cf. below).

The faunal analysis of the Möllbos community is based on more than 3400 valves, of which about 800 are articulated. Statistics suggest that the co-existence of the three non-siphonate deposit-feeders *Nuculodonta gotlandica, Nuculoidea lens,* and *Janeia silurica* (solemyoid, not included in Fig. 12) indicates niche diversification with feeding depths in order of increasing depth. In samples where the shallow burrower *Nuculodonta gotlandica* is numerically dominant over *Nuculoidea lens* and *Janeia silurica* taken together, the siphonates *Palaeostraba baltica* and *Caesariella lindensis* are significantly missing. Conversely, in samples where *Nuculoidea lens* and *Janeia silurica* together dominate over *Nuculodonta gotlandica*, the two mentioned siphonate species are most abundant, even though no more than 7.6%.

The siphonates *Palaeostraba baltica* and *Caesariella lindensis* probably co-existed with the non-siphonate *Nuculoidea lens* and *Janeia silurica* and inhabited different tiers, i.e. they occupied the shallowest level (Fig. 12:4, 5). *Nuculoidea lens* and *Janeia silurica* also show a significantly higher rate of articulated valves (33.8% and 31.0%, respectively) than does *Nuculodonta gotlandica* (17.9%) suggesting a deeper life position for the two first mentioned. *Nuculoidea* sp. A is found as one articulated specimen only, but because of its morphological features it most probably lived at the shallowest level (Fig. 12:3).

The unusually high number of fossil bivalves obtained from this fauna gives insight into other ecological phenomena. For example, a shell of *Nuculodonta gotlandica* shows pearl formation in the shape of a tube (Fig. 28G) representing a tumour formed by a possible commensal. Also blister-pearls have been recorded (28C; the oldest bivalve pearls known). An example of shell repair of the mantle edge is reflected in markings on the shell in one specimen of *Nuculodonta gotlandica* (Fig. 28A). Probably the damage was caused by a nibbling predator, but the mantle recovered gradually and eventually produced an almost normal shell margin. For a shorter period of time after the injury the valves grew at a slower pace, as reflected by change in shell shape. Growth-ring analysis of *Nuculoidea lens* indicates that the largest specimen (14 mm long) was about 7 years old (Liljedahl 1984a, p. 55).

The Djupvik fauna. – This fauna (Fig. 13) has not been bulk-sampled but is represented by specimens collected over a period of several decades. Only general synecological conclusions can therefore be drawn. Most specimens from Djupvik are articulated and preserved in marl indicating a biocoenosis. However, the specimens were most probably collected from a number of levels, and thus the relative numbers do not necessarily reflect the actual numerical relation between the different species at one time.

The nuculoid part of the fauna appears to have been dominated by non-siphonate forms, *Tancrediopsis gotlandica*, *Nuculoidea pinguis*, and *Praenucula faba* being the three most numerous. Altogether there are approximately 400–500 non-siphonate nuculoid specimens from this locality and about 100 siphonate ones, *Ekstadia tricarinata* being the most common of the siphonate species (95 specimens). Whether the siphonate *Ekstadia* was outcrowded by co-occurring non-siphonate species or formed a single-species community in beds separated from those with abundant non-siphonate species, cannot be determined from the present material.

The Burgsvik fauna. – The Burgsvik bivalve fauna (Fig. 14) is even more difficult to evaluate. The Burgsvik Beds form a complex series of different lithologies, ranging from coarse-grained conglomeratic limestones over oolitic limestones to pure sandstones. *Ledopsis burgsvikensis*, *Nuculoidea burgsvikensis*, and '*Nuculana*' *oolitica* are found in oolitic limestone while *Nuculites solida*, *Nuculites* sp. B, and ?*Nuculoidea ecau-*

data are found in limestone. As a whole, the Burgsvik Beds contain very few nuculoid specimens as compared with other bivalves. The oolitic limestone indicates a shallow environment with high water energy, but the fossils are usually well preserved, encrusted with a more or less thick coating of precipitated calcium carbonate. All specimens collected are single valves, which confirms the assumed high energy at the time of their accumulation.

Historical review

There exists a real difficulty in synthesizing the systematic data available on Palaeozoic nuculoid bivalves because most descriptions are dated and therefore difficult to use for comparative purposes. However, a thorough revision of the whole group is necessary, and as the revision work of the nuculoids proceeds, the often-used names of Recent taxa, such as, e.g., *Nucula* and *Nuculana* ('waste-paper basket' taxa for nuculoid-like bivalves throughout the fossil record), may eventually be abandoned and replaced by more appropriate names. Many of the generic and probably also of the specific names are synonyms, which makes a reliable compilation of taxa almost futile at present. Furthermore, in some cases the old literature is not specific as to stratigraphic level.

The earliest bivalve assumed to be a nuculoid or nuculoid progenitor is *Pojetaia* from the late Early Cambrian (Runnegar & Bentley 1983; Pojeta & Runnegar 1985). After its appearance there is a gap until the Ordovician, when 'true' nuculoids appeared.

The oldest post-Cambrian bivalves, which are nuculoids (palaeotaxodonts), have been reported from the Tremadocian of Argentina (Harrington 1938), Afghanistan (Termier & Termier 1971) and France (Thoral 1935). They are represented by four genera only, viz. *Ctenodonta*, *Palaeoneilo*, *Afghanodesma* and an un-named genus closely related to *Deceptrix* (Pojeta 1978, p. 230). In the Middle to Late Ordovician, nuculoids were represented by about 20 genera, and the period as a whole by almost 30 genera, making about $\frac{1}{3}$ of all known Ordovician bivalve genera (Pojeta 1978, p. 231).

Only about eleven of the Ordovician nuculoid genera survived into the Silurian, viz. *Cardiolaria*, *Cleidophorus*, *Ctenodonta*?, *Deceptrix*, *Nuculites*, *Nuculoidea*, *Palaeoneilo*, *Praenucula*, *Pyrenomoeus*, *Similodonta*, and *Tancrediopsis*. Ten exclusively Silurian genera have been reported, viz. *Arisaiga*, *Bicrenula*, *Caesariella*, *Dysodonta*, *Ekstadia*, *Gotodonta*, *Metapalaeoneilo*, *Nuculodonta*, *Palaeostraba*, and *Tropinucula*, as well as three additional genera which appeared during the Silurian and persisted into the Devonian, viz. *Ledopsis*, *Praectenodonta*, and *Tellinopsis*. A total of about 25 Silurian genera are represented, in Asia & Australia by about 10 species (of 6 genera), in North America by about 50 species (of 12 genera), in Bohemia by about 7–19 species (uncertain stratigraphical documentation) of 2–4 genera; in Great Britain by about 20 species (of 4 genera) and in Baltoscandia by about 25 species (of 11 genera).

Important works on Silurian bivalves treating Asia and Australia are those of Chapman (1908), Sherrard (1959), and Philip (1962); North America those of Simpson (1890), McLearn (1924), and Bambach (1969, unpublished); Bohemia that of Barrande (1881); Great Britain those of Sowerby (1839), McCoy (1862), Hind (1910), and Reed (1931); Baltoscandia those of Soot-Ryen (1964) and Liljedahl (1983, 1984a, b, 1985, 1986, 1989a, b, c, 1991, 1992a, b, c and this paper).

Modiomorphidae

Morphology

General shell shape. – The shell shape varies in lateral view from almost circular (Fig. 46B), to elongate subovate (Fig. 37J) to subtriangular (Fig. 39I), to elliptical (Fig. 37G). The shell may be equivalved (Fig. 37E) or subequivalved (Fig. 46C) and have a reduced anterior lobe (Fig. 37B) but generally without terminal beaks (Fig. 37B).

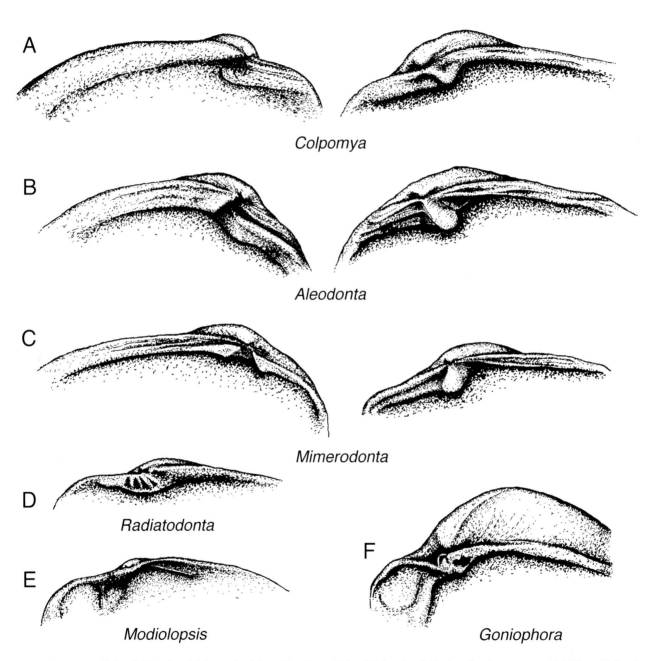

Fig. 16. □A. Hinge of left and right valve of *Colpomya hugini* n.sp. □B. Hinge of left and right valve of *Aleodonta burei* n.gen. et n.sp. □C. Hinge of left and right valve of *Mimerodonta atlei* n.gen. et n.sp. □D. Hinge of right valve of *Radiatodonta* sp. 1. □E. Hinge of right valve of ?*Modiolopsis alvae* n.sp. □F. Hinge of right valve of *Goniophora onyx* Liljedahl, 1984.

The beaks are prosogyrous (Fig. 37H) and inconspicuous (Fig. 45E), and in some species extremely small (Fig. 46H). They may even be involute (Fig. 53L).

In most species a more or less conspicuous umbonal ridge is present (Fig. 37J), sometimes developed as an angular keel (Fig. 53C). In some species there are two ridges, one anteriorly and the other posteriorly (Fig. 37K).

Shell sculpture. – The shell surface may have faint growth lines on a smooth surface (Fig. 37L), in some cases with distinct growth-increment stops (Fig. 46D), comarginal costellae (Fig, 54C), transverse costellae (Fig. 53O), or radial ribs (Fig. 54F).

Size. – The length of the largest specimens of the different species varies from about 13 mm (*Goniophora alei*) to about 65 mm (*Radiatodonta* sp. 1).

Byssal apparatus. – In byssate bivalves a byssal apparatus may be externally recognized by a more or less conspicuous byssal gape, a feature prominent in, e.g., North American Ordovician Ambonychiidae (Pojeta 1966, p. 141). Its presence may also be inferred by the existence of a byssal sinus, an indentation seen in lateral view in the anteroventral part of the shell, where the byssus threads protruded. A byssal gape is quite rare in Palaeozoic byssate forms (Pojeta 1966, p. 141), while a byssal sinus is a common feature in members of several groups of bivalves, e.g., Ambonychiidae, Modiomorphidae, and Pterineidae.

The presence of a byssal gape is difficult to establish in fossil material, especially in single valves. In some modern bivalves, e.g., *Mytilus* Linnaeus and *Modiolus* Lamarck, the location of the point where the byssus threads protrude, i.e. the gape, may not be readily distinguishable, owing to its minute size, in spite of the development of a strong byssus and, in some cases, also of a prominent byssal sinus.

Hinge. – North American Ordovician modiomorphids are either edentulous [(e.g., *Modiolopsis valida* Ulrich, *Modiolopsis* aff. *simulatrix* Ulrich, *Modiolopsis versaillensis* Miller, *Whiteavesia* cf. *cincinnatiensis* (Hall & Whitfield) and *Whiteavesia cincinnatiensis* (Hall), in Pojeta 1971, Pl. 14:1–5, Pl. 17:18–18, respectively)] or have only cardinal teeth (e.g., *Colpomya concentrica* Ulrich, in Pojeta 1971, Pl. 12:2–3, *Modiolodon oviformis* (Ulrich), in Pojeta 1971, Pl. 13:6,8–9 and *Modiomorpha concentrica* (Conrad), in Pojeta et al. 1986, Pl. 11:2,5, Pl. 12:1 and in Baily 1983, Fig. 47A, E).

The Swedish Silurian Modiomorphidae exhibit more variation (Fig. 16) in having a hinge (1) with a hinge plate that has a diagonal cardinal tooth in the right valve and a diagonal socket in the left (Fig. 16A); (2) with no hinge plate, a diagonal cardinal tooth in the right valve and a diagonal folding of the dorsal margin in the left (Fig. 16B); (3) with no hinge plate, an erect cardinal tooth in the right valve and a V-shaped slit in the dorsal margin of the left one (Fig. 16C); (4) with a hinge plate that has dorsally converging cardinal teeth and sockets in each valve (Fig. 16D); (5) with a hinge plate that has one or two diagonally arranged cardinal teeth and sockets in each valve (Fig. 16E); (6) with a hinge plate but without dentition (Fig. 16F).

Lateral dentition of the hinge is lacking in all the modiomorphid species studied.

Ligament. – Ligament structures of Palaeozoic bivalves are only rarely preserved. To my knowledge, only one account of preserved ligament material has been published, viz. the ligament of *Modiolopsis modiolaris* (Conrad) preserved as a dark stain (Pojeta 1971, Pl. 15:5). The ligament areas of fossil bivalves, however, are in many cases preserved well enough to provide information on the morphology of the original ligament.

The ligament of the present modiomorphids is opisthodetic, i.e. such that the mantle isthmus was situated posterior to the umbones (Trueman 1969, p. N60). It shows close resemblance to the ligament construction in living mytilaceans, by some authors referred to as parivincular or of the C-spring type (Dall 1895, p. 500; Newell 1942, p. 28; Trueman 1969, p. N61, Fig. 52c, D, D'; cf., however, Pojeta et al. 1986, p. 71, who consider the ligament of *Mytilus edulis* as not being of the C-spring type).

In the modiomorphids under study, as well as in other members of the group, there is a groove in the dorsal margin serving for the attachment of the ligament, [Figs. 17, 39B–D, F–H, K, 41B, E–G, 46A, E, L–M, 48B–D, H, 50D, F–G, I–K herein; cf. also Bailey 1983, Fig. 47E; and Pojeta et al. 1986, Pl. 11:5 of *Modiomorpha concentrica* (Conrad); Pojeta et al. 1986, Pl. 14:2, 4 of *Modiolopsis* aff. *simulatrix* Ulrich, and Pl. 14:5 of *Modiomorpha versaillensis* Miller)]. Most probably, it corresponds in function to the resilial ridges of modern mytilids.

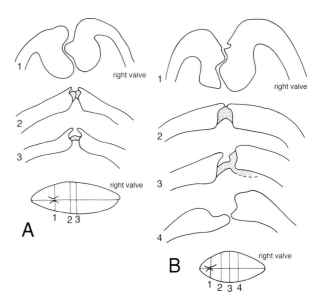

Fig. 17. Camera lucida drawings of cross sections through hinge and ligament regions. Ligament remnants stippled. □A. *Colpomya hugini* n.sp. □B. *Aleodonta burei* n.gen et n.sp.

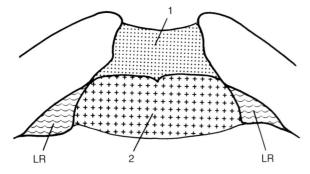

Fig. 18. Cross section of ligament region of *Mytilus edulis* showing shell (blank), ligament ridges (LR), outer layer of ligament (1), and inner layer of ligament (2). (After Trueman 1950, Text-fig. 3.)

In some of the material at hand, cross sections show remains of the ligament proper. Fig. 17A:a is a cross section of the hinge region, and Fig. 17A:b–c are of the middle part of the ligament of *Colpomya hugini*. Fig. 17B:a is a cross section of the hinge region and 17B–C of the ligament, and 17D of the shell posterior to the ligament of *Aleodonta burei*.

Since the inner layer of the bivalve ligament is calcified (the periostracum and outer layer are not; Trueman 1969, p. 58), it is more prone to fossil preservation than the uncalcified tissues of the ligament. Thus, the ligament structures preserved in *Colpomya hugini* and *Aleodonta burei* probably correspond to the inner ligament layer of, e.g., *Mytilus edulis* (layer 2 in Fig. 18).

Musculature. – The present modiomorphids are anisomyarian, that is, the anterior adductor muscle scar is reduced in size relative to the posterior one (Fig. 19B) The anterior adductor muscle scar is generally more deeply impressed (Fig. 46I, K) and in some cases extremely so (Figs. 37E, 43F–G).

In some of the Silurian modiomorphids, both the anterior and posterior adductor muscle scars show evidence of 'quick' and 'catch' structures (Fig. 19A, C, F). The 'quick' muscle fibers of bivalves are striated and believed to function during rapid closing of the valves, whereas the 'catch' muscle fibres are smooth and work more slowly, keeping the valves closed, in some cases for a considerable time (Cox 1969, p. N35). In dimyarian bivalves the 'quick' muscles occupy the part of the scar close to the middle of the shell, while the 'catch' portion is located in the peripheral part of the scar (Cox 1969).

In those species of the present material where 'quick' and 'catch' portions are discernible, the area closest to the middle of the shell is typically larger than the peripheral part. If the distribution of the two kinds of muscles of the Silurian modiomorphids was similar to that in living bivalves, it appears that the 'quick' portion of at least some of the present modiomorphids is larger than the 'catch' part. The need of frequent and rapid closure of the valves in non-swimming bivalves may be required for two main resons, viz. for cleans-

ing the mantle cavity from undigestible or foreign matter and for protection against physical disturbances, such as, e.g., predation.

The fact that the 'quick' portion of the adductor muscle scar dominates over the 'catch' portion, suggests ability of frequent and rapid closing of the valves. As predation pressure on Silurian bivalves was probably low (Liljedahl, 1985), compared to that in modern biota, it is likely that the ability to close the valves rapidly was primarily required for the cleansing action of removing undigested and foreign matter from the mantle cavity. Probably this cleansing method was prevalent in Silurian times, owing to the fact that siphons and mantle fusion apparently were only extremely rarely developed in Palaeozoic suspension-feeding bivalves.

The anterior adductor muscle scar of some of the present modiomorphids exhibit an uneven, jagged posterior border (Fig. 19D, 23H) suggesting the presence of several accessory muscles (cf., e.g., such muscles in *Tellina*, Cox 1969, p. N30, Fig. 31H). Also, more or less conspicuous single scars in the anterior region of the shell may be present (Fig. 19A). Commonly, an extraordinarily deep, relatively small scar is situated between the anterior adductor muscle scar and the umbonal cavity, that reflects the impression of a strong pedal–byssal retractor muscle (Figs. 19A, 37A, C, 43C, F, G, 45H, 46I, 55C–D, F). Some species exhibit one or more impressions in the summit of the umbonal cavity, indicating the presence of pedal elevators and possibly also of visceral muscles (Figs. 19A, 37A, C).

In some species the posterior adductor muscle scar shows hypertrophism (Figs. 19C, E–F, 37F, 39I, 45H, 55A). In Recent Mytilacea the posterior adductor muscle is connected with anterodorsally extended bundles of the posterior byssal

Fig. 19. Muscular impressions. ☐A. *Modiodonta gothlandica* Liljedahl, 1989. Anterior end of internal mould of articulated specicmen showing 'quick' and 'catch' portions of the anterior adductor muscle scar and pedal–byssal scars (for explanation see below). RMMo 24944, ×12.3. ☐B. *Colpomya hugini* n.sp. Internal mould of right valve showing anisomyarian musculature of articulated specimen. RMMo 25420, ×6.4. ☐C. ?*Colpomya balderi* n.sp. Posterior end of articulated specimen showing 'quick' and 'catch' portions of the posterior adductor muscle scar, and secondary line of pallial attachment (at arrows). RMMo 150416, ×3.0. ☐D. *Aleodonta burei* n.gen. et n.sp. Anterior part of internal mould of articulated specimen showing anterior adductor muscle scar with jagged posterior limitation (arrows) indicating the positions of accessory muscles. RMMo 25428, ×7.0. ☐E. *Colpomya burei* n.sp. Internal mould of articulated specimen showing hypertrophied dorsal part of posterior adductor muscle. RMMo 16523, ×8.9. ☐F. *Modiodonta gothlandica.* Internal mould of articulated specimen showing scar of hypertrophied posterior adductor muscle and 'quick' and 'catch' portions of this scar. RMMo 24942, ×5.0. All specimens from the Wenlockian Mulde Beds at Djupvik, Gotland. aa = anterior adductor, pa = posterior adductor, q = quick portion, c = catch portion, pe = pedal elevator, pp = pedal protractor, pr = pedal retractor, pbr = pedal–byssal retractor.

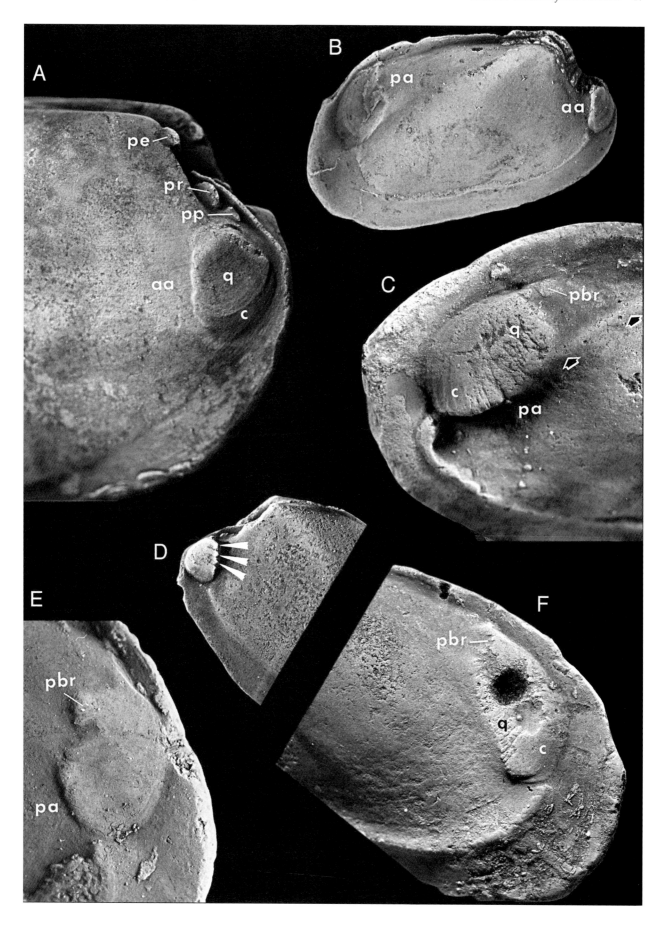

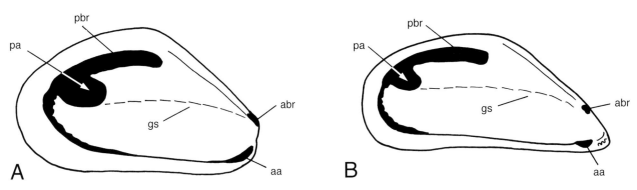

Fig. 20. Musculature of □A. *Modiolus modiolus.* □B. *Mytilus edulis.* a*a* = anterior adductor muscle, abr = anterior byssal retractor muscle, gs = gill suspensories, p*a* = posterior adductor muscle, pbr = posterior byssal retractor muscle. (After Newell 1942, Fig. 6A, D.)

and pedal retractor muscles (Fig. 20). Thus, the impressions in question of the modiomorphids under study most probably show the positions of such muscles. Also, additional small muscle imprints, referable to pedal muscles, are present in the posterior region (Fig. 46I).

Pallial muscles of the mantle are often present as radial or punctiform scars connecting the adductor muscle scars. In one species, ?*Colpomya ranae*, the pallial line is slightly sinuated (Fig. 43H). In another, ?*Colpomya balderi*, the pallial line continues on the dorsal part of the posterior adductor muscle scar (Fig. 19C, 45H; cf. secondary line of pallial attachment of *Mytilus in* Owen 1958, p. 642, Fig. 4A; see also discussion in 'Classification').

Some of the present species, *Modiodonta gothlandica* (Fig. 37G), *Colpomya hugini* and ?*Colpomya ranae*, (Fig. 43G–H), exhibit internal diagonal ridges, possibly being the attachment area of the gills (cf. similar structures in the myalinids *Myalina goldfussiana* and *Orthomyalina slocomi*, and also gill suspensories, in the Recent *Modiolus modiolus* and *Mytilus edulis, in* Newell 1942, p. 30, Fig. 6E, G, A, and D, respectively).

Functional morphology, ecology, and life habits

Shell morphology of bivalves is influenced by environmental factors and may be conspicuously similar in distantly related bivalves occupying the same environmental niche. In numerous different phylogenetic groups of bivalves, great morphological diversity may occur because of adaptions to a wide range of habitats.

Epifaunal groups are more variable in morphology than infaunal groups (Kauffman 1969, p. N142). For the latter ones, shape and relative convexity are closely related to habitat depth in the sediment and rate of burrowing (Kauffman 1969, pp. N142, 166).

Shell thickness, for instance, may be related to burrowing rate, in that thick-shelled species normally are slow burrowers and thin-shelled ones rapid (Stanley 1970, p. 61). Rapid

burrowers are either deposit feeders or inhabitants of moderately unstable or unstable shifting substrates (Stanley 1970, p. 61).

Burrowing rate can also be related to shell ornamentation in that strongly ornamented forms are normally slow burrowers, while ornamentation is weak or absent in most rapid burrowers. However, some slow burrowers have little or no ornamentation (Stanley 1970, p. 62).

In the living relatives of Modiomorphidae, the Mytilidae, there are two major life habit groups, endobyssate semi-infaunal or epibyssate epifaunal species (Stanley 1972, p. 169). They are all characterized by adult attachment, a reduced anterior region, and an anisomyarian or monomyarian musculature (Stanley 1972, p. 169). The endobyssate species exhibit moderate reduction of the anterior lobe and have a more cylindrical shell form than epibyssate ones. Epibyssate species, generally lacking the anterior lobe, have a flattened venter and only a diminutive anterior adductor muscle.

In cross section the maximum convexity of endobyssate bivalves is at or above mid-height of the shell, while in epifaunal forms the maximum convexity is below mid-height (Fig. 21A). The musculature of both endobyssate and epibyssate species is anisomyarian, i.e. the anterior adductor muscle is reduced in size (in endobyssate species there is only moderate reduction), and in some epibyssate the anterior adductor muscle may even be entirely lacking, owing to extreme reduction of the anterior part of the shell.

In endobyssate species a line trough the centres of the adductor muscles intersects with the hinge axis at an acute angle (posterior auricle developed), and in epibyssate species the lines intersect at an obtuse angle (posterior auricle strongly developed; Fig. 21B). In the latter the posterior byssal retractor muscle is situated immediately above the byssus in a dorsal position, thus providing a powerful attachment, while endobyssate forms have their byssal retractor in a more posterior position resulting in a weaker byssal attachment (Fig. 21B: see Stanley 1972, Fig. 3D).

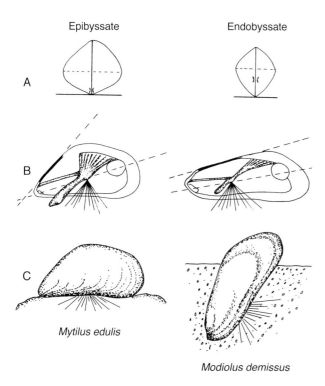

Fig. 21. Comparative adaptive morphology of *Mytilus edulis* (epifaunal) and *Modiolus demissus* (semi-infaunal). □A. Cross-sectional shapes showing position of maximum width in relation to height of the shell. □B. Internal view showing musculature and relation between the lines passing through ligament and adductor muscle scar (see text for explanation). □C. Life positions on hard surface and in soft substrate, respectively (after Stanley 1972, Text-fig. 3).

In several species of the genus *Modiolus*, a twisted commissure plane is a common feature (Savazzi 1984, p. 307). In these cases the animal lives semi-infaunally, being attached with a fairly strong byssus, and resting on one valve, i.e. with the commissure plane inclined. Thus, shell torsion reduces the height and vulnerability of the parts projecting above the sediment surface, but at the same time it maximizes the length of the posterior commissure raised above the sediment surface (Fig. 22).

The posterior adductor muscle of isofilibranchs is often connected with a variable number of posterior byssal retractor muscles. In the anterior part of the shell, close to the adductor muscle, there is also a deeply impressed byssal retractor muscle, which in some species is attached to an umbonal septum. The pallial line in living mytilids is integripalliate but often discontinuous between the impressions of the two adductor muscles.

Shell shape of the taxa under study, except *Goniophora*, (see separate discussion below), varies from rounded oval to obliquely elongate to almost mytiliform (length/height ratio 1.1–2.5). The valves are more or less compressed and generally have a smooth surface. The length/height ratio to height/width ratio of the species are presented in Fig. 23. There are no extreme shape types among the species studied, such as disc-, sphere-, blade- or cylinder-shape. Most forms are more or less close to the line separating the area of rapid burrowing from that of non-rapid burrowing in Fig. 23. The most blade-

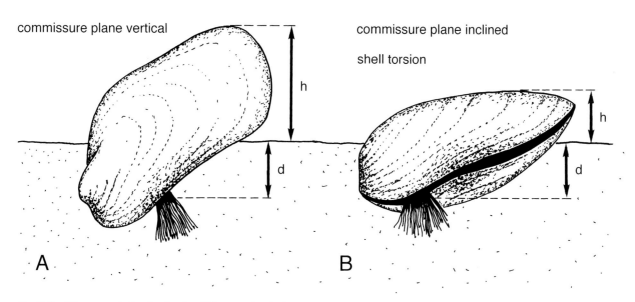

Fig. 22. Semidiagrammatic drawing showing difference in height above the sediment surface between shell with commissure plane vertical (A) and inclined with shell torsion (B). h = height above sediment surface, d = byssal depth (after Savazzi 1984).

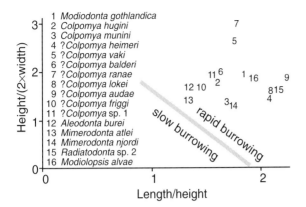

Fig. 23. Shell shape related to burrowing rate according to Stanley's (1972, Fig. 25) scheme.

like forms are ?*Colpomya ranae* and ?*Colpomya vaki* (Nos. 7 and 5, respectively, in Fig. 23) while *Mimerodonta atlei* (No. 13 in Fig. 23), ?*Colpomya friggi* (No. 10 in Fig. 23) and *Aleodonta burei* (No. 12 in Fig. 23) approach sphere shape, although they are still far from being completely spherical.

Although shell thickness has not been measured, it appears as if the Swedish Silurian modiomorphids had comparatively thin shells. ?*Colpomya ranae*, *Modiodonta gothlandica*, ?*Colpomya vaki*, ?*Colpomya audae*, ?*Colpomya lokei*, and *Modiolopsis alvae* all have comparatively thin shells, while *Aleodonta burei* and *Mimerodonta njordi* have moderately thin shells. ?*Colpomya heimeri*, ?*Colpomya friggi*, ?*Colpomya* sp. 1, *Radiatodonta* sp. 1, and *Radiatodonta* sp. 2 seem to have had a relatively thick shell.

Accordingly, the elongate, compressed and thin, smooth shell of ?*Colpomya ranae*, together with the configuration of the muscular imprints, suggests a rapid-burrowing, possibly completely infaunal (suggestion based on the presence of a pallial sinus), non-byssate life habit. This assumption is further strengthened by the presence of distinct radial attachment scars of the mantle, which possibly indicates fused mantle margins typical of rapid burrowing siphonate species. In *Modiodonta gothlandica*, *Modiolopsis alvae*, ?*Colpomya balderi*, and *Aleodonta burei* the detailed morphology of the muscular impressions supports the assumption of ability of relatively fast burrowing and a moderately strong byssal attachment in a semi-infaunal position.

In order to explain the functional morphology of the moderately thin-shelled species, one has to consider more factors than burrowing ability. Shell thickening in living bivalves increases physical stability (Stanley 1970, p. 82). In unstable substrates, however, a thin shell combined with rapid burrowing are the best solutions to the problems of adapting to this kind of habitat. Some rapid burrowers, however, have thick shells which counteract disinterment by increasing whole-animal density (Stanley 1970, p. 82).

The moderately thin-shelled modiomorphids of this paper, except for *Aleodonta burei* (see below), have been found

in coarse-grained sediments, and most of them (except for *Goniophora*) are considered relatively fast burrowers (see, however, *Aleodonta burei*, *Mimerodonta atlei* and *Mimerodonta njordi* below).

Two of the species studied, *Colpomya hugini* and *Colpomya munini*, are slightly inequivalved as observed in a large number of articulated specimens. This was possibly also the case in *Aleodonta burei*, *Mimerodonta atlei*, and *Mimerodonta njordi*. Although the differences in morphology of the valves of these species in some cases is slight, there probably was a functional significance. In infaunal as well as in deeply semi-infaunal bivalves, such differences are not likely to have been of functional importance. Instead, if an inequivalved bivalve has a shallowly semi-infaunal, low-angle fixation having the valve with the more developed umbonal ridge oriented upwards, this valve may have channelled the water flow more efficiently than would have been the case with the other valve up (cf. endo-byssate, low angle fixation suggested for Palaeozoic pterineids in Stanley 1972, p. 185). The presence of an anterior lobe in the modiomorphids studied does not, however, preclude an epifaunal life habit, since living mytilids with a pronounced anterior lobe may be fully epifaunal [e.g., *Adipicola pelagica* (Woodward, 1854) *in* Dell 1987, Fig. 1–2, *Adipicola simpsoni* (Marshall, 1900) *in* Dell 1987, Figs. 5–6, *Benthomodiolus abyssicola* (Knudsen, 1970) *in* Dell 1987, Figs. 46–47)].

Slightly inequivalved shells are unsuitable for effective burrowing. In most byssally attached anomids and adult pectinids the less convex valve is usually facing the substratum for increased stability (Stanley 1972, p. 30). Stanley (1972, p. 185) suggested that a minor reduction in the right valve convexity of endobyssate pterineids was associated with a low-angle fixation for increased stability in a less deeply buried life position. Thus, the shell shape of *Aleodonta burei* suggests – together with the configuration of its muscular imprints and the presence of epibionts on the shell – ability of slow burrowing and a shallow, endo-byssate life position, possibly in a low angle fixation (Fig. 47B).

Another life habit, present in some Recent mytilids, is the nestling in the byssus threads of the bivalve individuals (e.g., *Musculus koreanus* Ockelmann, 1983, p. 96). However, the hard part morphology alone does not give any information as to this kind of life habit.

At least one of the modiomorphid species studied, viz. *Colpomya munini*, exhibits a twisted commissure plane (Fig. 41D). Besides the fact that this species is inequivalved, its shell torsion suggests that it was living infaunally, with the plane of commissure inclined, in a low-angle fixation (Fig. 25).

The inequivalved condition has been recognized in, e.g., the Myalinidae, as intermediate between inequivalved infaunal and semi-infaunal species and equivalved forms. Early Palaeozoic semi-infaunal, modiomorphid-like forms may have been forerunners of late Palaeozoic, epifaunal mytilids (Stanley 1972, p. 178, Text-fig. 12). The inequivalve, endobyssate modiomorphids studied possibly represent an early attempt to exploit a niche more fully exploited by later bivalve groups.

The majority of the present species exhibits morphological features typical of extant endo-byssate taxa, such as elongate shell form, maximum convexity above mid-height, reduced but lobate anterior end, and, in some cases, both anterior and posterior byssal retractor muscle scars. Extant endo-byssate mytilids have a reduced foot and are not true burrowers (Stanley 1972, p. 117). However, the species studied also show similarities to active burrowers, primarily the compressed shell shape and the configuration of the muscular imprints. The reduction of the anterior adductor muscle scar is not as pronounced as in living mytilids (active burrowers are iso-myarian). In several species, scars in the umbonal cavity are present, and these are interpreted as the impressions of pedal elevators. Such muscles are lacking in mytilids, but they are present in active extant burrowers such as *Tellina, Anodonta,* and *Neotrigonia.* In a number of the present Silurian species, the anterior adductor muscle scar has an uneven, jagged posterior edge indicating the presence of pedal retractors and protractors (cf. *Tellina, Anodonta,*) The position of the posterior pedal–byssal retractor muscle scars of the present modiomorphids suggests a less powerful byssal attachment than in living mytilids (cf. highly developed posterior byssal and pedal retractors in *Modiolus* and *Mytilus* in Newell 1942, p. 30, Fig. 6A, D).

The genus *Goniophora* exhibits a variety of shell forms indicating different modes of life. All species are equivalved and characterized by, among other things, their diagonal umbonal keel. The keel may be almost straight, seen in lateral view, as in *Goniophora acuta* (Fig. 54B, D). This species has flattened valves as seen in dorso-ventral direction (width/length value less than 0.5), i.e. the animal is broad and flat. The keel may also be more or less sigmoidal, as in the remaining *Goniophora* species. The species with the most sigmoidal keel is *Goniophora alei* (Fig. 53K), which also has the highest width/length ratio (0.84) and a comparatively high shell.

Some species, such as *Goniophora onyx, Goniophora brimeri, Goniophora bragei,* and *Goniophora tyri,* have a moderately reduced anterior lobe. '*Goniophora*' sp. 1 and '*Goniophora*' *gymeri* have a strongly reduced anterior lobe, while *Goniophora acuta* exhibits a complete loss of the anterior lobe.

Goniophora chemungensis (Vanuxem), which belongs to the group with moderately sigmoidal keel and slight reduction of the anterior lobe, was suggested to have had a semi-infaunal endobyssate life habit by Stanley (1972, p. 194, Text-fig. 28F). He illustrated it with only the posterior margin protruding. Bowen *et al.* (1974, p. 107, Fig. 15:9) presented a similar interpretation of the life position of *Goniophora* in general. Drevermann (1902, Pl. 11:4), however, figured an *in situ* specimen of *Goniophora stürtzi* Beushausen having a slightly sigmoidal keel and moderately reduced anterior lobe, with the diagonal keels parallel to the bedding plane.

A shell form with a straight umbonal keel, and a broad, flat shell, such as that in *Goniophora acuta,* has a large surface area relative to total volume (body mass). Owing to the so called snowshoe effect (Thayer 1974), an animal with this shape is able to maintain a stable life position on a fine-grained, muddy substrate. In the case of *Goniophora acuta* the shell is furthermore furnished with a fine concentric sculpture, as well as an additional coarser radial pattern on the ventral part of the valves. Its almost completely reduced anterior lobe, convex ventral margin, and low posterior profile indicate absence of a functional foot as well as a byssus. Foot and byssus are unnecessary in a stable, low-energy environment, which was probably the habitat of *Goniophora acuta,* as indicated by the fine-grained, argillaceous sediment in which it has been found.

The *Goniophora* species with moderate sigmoidal keels and moderate reduction of the anterior lobe are intepreted as having had a life position with the keels parallel to the sediment surface (snowshoe effect), but in a more semi-infaunal manner, since the anterior lobe would dip down into the substrate (e.g., *Goniophora onyx* in Fig. 24). The shape of

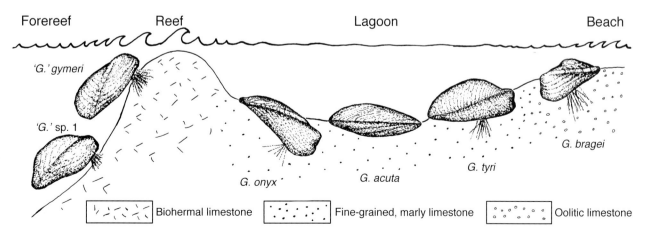

Fig. 24. Suggested life position and tiering of some *Goniophora* species in a generalized section of the Silurian coast of Gotland.

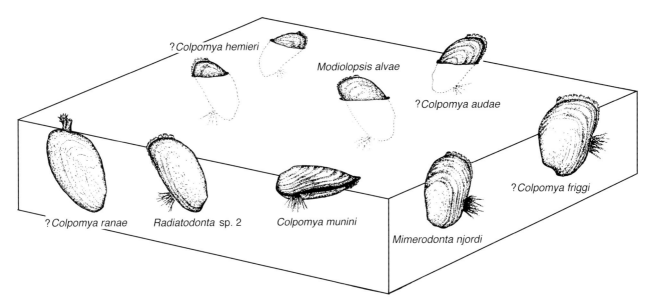

Fig. 25. Reconstructed life position of some of the modiomorphids from the Ludlovian Burgsvik Beds of Gotland.

the anterior lobe and the presence of a pedal–byssal muscle scar in *Goniophora brimeri* suggest that this species had a byssus, and because of their similar shell morphology, *Goniophora onyx*, *Goniophora bragei*, and *Goniophora tyri* most probably also had a byssus.

Goniophora alei, '*Goniophora*' *gymeri*, and '*Goniophora*' sp. 1 are in some respects (e.g., terminal beak, straight ventral margin and extremely thick and strong hinge region, Figs. 53M, 54L) reminiscent of the unrelated epifaunal ambonychiids. The umbonal keels of these '*Goniophora*' species are highly sigmoidal, and the beak of species of this group is almost terminal (Figs. 53N, 54I, M). Thus, the anterior lobe is much reduced, and a hypothetical line trough the adductor muscle scars intersects the hinge axis at a wide angle, which is the case also in living epibyssate mytilids (Stanley 1972, p. 171, Text-fig. 3). Extant bivalves with anterior reduction and expanded posterior end have an epibyssate life habit (Stanley 1972, p. 173).

'*Goniophora*' *gymeri* shows close external resemblance to several species of the Pennsylvanian *Promytilus* Newell, e.g., *Promytilus annosus annosus* Newell and *Promytilus vetulus* Newell (Newell 1942, Pl. 1:9, 3). Representatives of this genus are believed to have lived in the littoral or sublittoral zone attached to a hard substrate (Newell 1942, p. 39). Stanley (1972, p. 175) compared the morphology of *Promytilus* with that of the Recent '*Modiolus*' *pulex* which lives attached to rocky intertidal surfaces. Thus, '*Goniophora*' *gymeri*, and also the closely similar '*Goniophora*' sp. 1, may have had an epifaunal mode of life (Fig. 24).

Thus, four modes of life are recognized for the modiomorphids studied: (1) epibyssate, that is, epifaunally with byssus threads attached to a firm substratum, (2) endobyssate, semi-infaunally attached with a byssus, (3) non-byssate semi-infaunally buried, and (4) non-byssate infaunally, completely buried in the sediment, and possibly using siphons for respiration (Figs. 24–25).

Historical review

One of the earliest records of Silurian modiomorphids is that of Hisinger (1827). He listed two species of the genus *Mytilus* Linnaeus (given, however, as *Mytilites* in his table on page 323), one of which is *Modiodonta gothlandica* (Hisinger) from the Wenlockian of Gotland, Sweden. Later, he (Hisinger 1831a, b) referred to this species as *Modiola gothlandica* Hisinger and still later, in 1837, he changed the name to *Modiola*? *nilssoni* Hisinger. The valid name of this taxon is *Modiodonta gothlandica* (Hisinger), as discussed by Liljedahl (1989, p. 313). No internal features of this species were recorded by Hisinger.

In 1839 Sowerby described external features of the Wenlockian species *Modiola antiqua* Sowerby and the Ludlovian *Modiola*? *semisulcata* Sowerby, and the interior of the Pridolian *Goniophora cymbaeformis* Sowerby, all from Britain.

The German modiomorphids described by Goldfuss (1836–1840) are difficult to evaluate, both systematically and stratigraphically.

Portlock (1843) described some British Silurian modiomorphids, viz. *Mytilus cintus* Portlock, *Mytilus*? *semirugatus* Portlock, *Modiola brycei* Portlock, *Modiolopsis expansa* Portlock, *Modiolopsis nerei* (Münster), and *Modiolopsis securiformis* Portlock. No internal characters were recorded.

Phillips & Salter (1848) considered *Modiolopsis* Hall to be a synonym of *Mytilus* Linnaeus and recorded the following species from the Silurian of Britain: *Mytilus gradatus* Salter, *M. perovalis* Salter, *M. platyphyllus* Salter, *M. exasperatus*

Phillips, *M. mytilimeris* Conrad, *M. chemungensis* Conrad, and *M.? ungiculatus* Salter. No internal characters were given.

In 1853 Ribeiro & Sharpe reported *Modiolopsis elegantulus* Sharpe from the Silurian of Portugal. No internal features were recorded.

From the Silurian of Britain McCoy (1855) described *Modiolopsis complanata* (Sowerby), *Modiolopsis nilssoni* (Hisinger), *Modiolopsis postlineata* (McCoy), *Modiolopsis solenoides* (Sowerby), and *Orthonotus cymbaeformis* (Sowerby). Only external features were accounted for.

Among the Silurian species from U.S.S.R. referred to as modiomorphids by Eichwald (1860), *Modiolopsis decussata* seems to belong to the genus *Rhombopteria*, *Modiolopsis complanata* is not figured, and *Mytilus uncinatus* appears to be an ambonychiid. No internal features were observed.

Barrande (1881) figured 40 new modiomorphids from Bohemia. Seven of the species are Late Ordovician or Early Silurian in age, and 26 of Late Silurian or Early Devonian age (see Kříž & Pojeta 1974 and discussion in 'Early modiomorphid faunas' herein). Only 11 species are of certain Silurian age, viz. *Modiolopsis concurs*, *M. involuta*, *M. propinqua*, *M. pupa*, *M. rebellis*, *M. senilis*, *M. tenera*, *Goniophora phrygia*, *G. retrorsa*, *G. scalena*, and *G. zephyrina*. Barrande's illustrations show the external appearances of the specimens only. I have studied his types, and only a fraction of the specimens exhibits hinge features. Most are impressions or internal moulds and badly preserved, and thus they are of little systematic value.

From the Silurian of Australia, Chapman (1908) described *Modiolopsis melbournensis* Chapman, *M. complanata* Sowerby, *M. nasuta* var. *australis* Chapman, *Goniophora australis* Chapman, and *G.* cf. *glaucus* Hall. No internal features were given.

Williams (1913) described *Modiolopsis leightoni* Williams and *M. leightoni* var. *quadrata* Williams and also illustrated *Eurymyella shaleri* var. *minor* Williams, all from the Silurian of North America. No internal features were observed.

The Silurian bivalves of Arisaig, Nova Scotia, were described thoroughly by McLearn (1924), but no hinge structures of the modiomorphids were observed. The following species were included: *Modiolopsis haliburtoni* McLearn, *M. exilis* Billings, *M. rhomboidea* Hall, *M. rhomboidea* var. *eurymyellaformis* McLearn, *M. rhomboidea* var. *subnasuta* (Hall), *M. rossonia* McLearn, *M. latouri* McLearn, *M.* sp., *Goniophora consimilis* Billings, *G. mediocris* Billings, *G. laticostata* McLearn, *G. transiens* Billings, *G. montsi* McLearn, *Cosmogoniophora bellula* (Billings), *G. bellula* var. *elongata* McLearn, and *Orthodesma?* sp.

Twenhofel (1927) described *Modiolopsis miser* Twenhofel from the Silurian of Anticosti Island (Canada) without giving any internal characters.

In 1927 Reed described the following species from the Silurian of Britain. *Goniophora (Cosmogoniophora) decorata* Reed, *Modiolopsis (Colpomya) concors* Reed (hinge described), and *M.? solenoides* (Sowerby).

In 1939 Northrope described the external characteristics of *Modiolopsis rhomboidea* Hall, *M. rhomboidea eurymyellaformis* McLearn, *M.* cf. *rhomboidea* Hall, *M. leightoni quadrata* Williams, *M. exilis* Billings, *M. perlata* Hall, *M. subcarinata* Hall, *Goniophora consimilis* Billings, and *Eurymyella shaleri* Williams, all from the Silurian of North America.

Sherrard (1959) described from the Silurian of Australia *Modiolopsis elongata* Sherrard and *Cosmogoniophora sinuosa* Sherrard. No hinge characters were accounted for.

Liljedahl (1984) described in detail the external and internal features of *Goniophora onyx* Liljedahl from the Wenlockian of Gotland, Sweden. In 1989 Liljedahl described the external and internal characteristics of *Modiodonta gothlandica* (Hisinger, 1831).

Thus, prior to the present paper, only three modiomorphids of Silurian age were to my knowledge well known in terms of the hinge, viz. *Modiolopsis (Colpomya) concors* Reed, *Goniophora onyx* Liljedahl, and *Modiodonta gothlandica* (Hisinger). The remaining species were unknown as to hinge features and accordingly difficult to evaluate systematically.

Early modiomorphid faunas

The oldest isofilibranch known is *Fordilla troyensis* Barrande from the middle and late Early Cambrian (Atdabanian–Botomian). Among bivalves it is preceeded only by the nuculoid progenitor *Pojetaia runnegari*, which appeared already in the Tommotian (Runnegar & Bentley 1983, p. 76).

After the appearance of these early bivalves there is a general gap in the stratigraphical distribution of the class until the Early Ordovician, when 'true' nuculoids and modiomorphids appeared (Pojeta & Runnegar 1985, p. 307).

The oldest bivalves referred to modiomorphid taxa seem to be *Modiolopsis ramseyensis* Hicks, 1873, *M. homfrayi* Hicks, 1873, *M. solvensis* Hicks, 1873, and *M. cambriensis* Hicks, 1873, from the Tremadoc of Wales (Hicks 1873, Pl. 5:14–20). Only the external features of these forms are known, and their generic affiliations are therefore difficult to evaluate. However, *Modiolopsis ramseyensis* resembles *Xestoconcha kraciukae* Pojeta & Gilbert-Tomlinson, 1977 from the Arenig of Australia, in which the hinge characters are known. *M. ramseyensis* may thus be a true modiomorphid. However, the shell shape of these two species is fairly common in bivalves and may be recognized in other groups as well (cf., e.g., the nuculoid *Ctenodonta primaeva* Beushausen, 1895, Pl. 5:28).

Although its internal features are unknown, *Modiolopsis davyi* Barrois, 1891, Pl. 3:7 (also figured by Babin 1966, Pl. 7:13) from the Arenig of France may be one of the earliest representatives of the group. *Xestoconcha kraciukae* Pojeta & Gilbert-Tomlinson, 1977, and *Colpantyx wolleyi* Pojeta & Gilbert-Tomlinson, 1977, from the Arenig of Australia are the oldest species with certainty recognized as modiomorphids. From the Ordovician are reported about 40 modiomorphid genera including more than 300 species (Vokes, 1980; Pojeta, 1971).

Prior to the present study the following generic names of Silurian modiomorphids have been used: *Mytilus* Linnaeus, 1758, *Modiola* Lamarck, 1801, *Modiolopsis* Hall, 1847, *Goniophora* Phillips, 1848, *Orthodesma* Hall & Whitfield, 1875, *Colpomya* Ulrich 1893, *Eurymyella* Williams, 1912, *Cosmogoniophora* McLearn, 1918, and *Modiodonta* Liljedahl, 1989b.

About 45 Silurian species have been assigned to *Modiolopsis* (27 in Europe, 13 in North America and 5 in Australia). Some additional 17 species assigned to this genus may be of Silurian or entirely Ordovician or Devonian age. This is due to the fact that Barrande (1881) reported *Modiolopsis* from the subdivision d5 of étage D (in which subdivision fossils of Late Ordovician to Early Silurian occur) to e2 of étage D (in which fossils of Late Silurian to Early Devonian occur; see Kříž & Pojeta 1974, p. 491, table 1). *Modiolopsis* is a 'waste-paper basket' taxon for numerous modiomorphid-like forms (see, e.g., Pojeta 1971, p. 7, reporting some 163 species assigned to this genus!); *Modiola* (e.g., *in* Hisinger 1831a, Sowerby 1839) and *Mytilus* (*in* Hisinger 1827) are used in the same way.

About 13 species have been assigned to *Goniophora* [also a 'waste-paper basket' genus (5 in Europe, 6 in North America and 2 in Australia)] and 12 additional species may be included in the European faunas (see discussion above on Barrande's stratigraphy concept).

One North American and one European species have been referred to the genera *Cosmogoniophora* (synonym of *Goniophora*) and *Colpomya*, and one species each to the genera *Eyrymyella* (North America), *Orthodesma* (North America), and *Modiodonta* (Europe).

At present the following genera have been recognized in European Silurian deposits: *Modiodonta*, *Colpomya*, *Cosmogoniophora*, *Goniophora*, *Modiolopsis*, *Aleodonta*, *Mimerodonta*, and *Radiatodonta*; in North American Silurian rocks *Eurymyella*, *Orthodesma*, *Colpomya*, *Cosmogoniophora*, *Modiolopsis*, and *Goniophora*; and in Australian rocks of corresponding age *Modiolopsis* and *Goniophora*.

From the Silurian of Europe two species of North American Devonian modiomorphids have been reported, viz. *Mytilus chemungensis* Conrad and *M. mytilimeris* Conrad. From the Silurian of Australia are reported the European Silurian *Modiolopsis complanata* Sowerby, and the Devonian North American *Goniophora* cf. *glaucus* Hall.

With the fact in mind that most determinations of *Modiolopsis* and *Goniophora* have been made on external features alone, and accordingly these two names have been badly abused, a compilation like this may be of little value. Furthermore, several of the modiomorphid taxa studied herein are difficult to determine as to generic affiliation, thus, systematic comparisons with taxa of other regions may be even more hazardous.

Faunal affinities

The different species studied occur in a wide range of lithologies. For instance, the great morphological variation within *Goniophora* may be explained as adaptations to a variety of distinctly different habitats. Fig. 24 shows the position of some of the *Goniophora* species in terms of their preferred habitats in a generalized shelf transect on Gotland during late Silurian time.

Littoral or sublittoral environments are represented by the Burgsvik sandstones and oolites. *Goniophora bragei* has been recorded in both these rock types, while *Goniophora alei* has been found only in the oolites. Both these species are characterized by a small, broad shell which was attached with a byssus. *Goniophora bragei* thus could maintain a stable, semi-infaunal life position in such unstable, shifting environments.

Fine-grained sediments, such as siltstones and marly limestones, are typical of back-reef or lagoon environments. In these we may find the medium-sized *Goniophora brimeri*, *G. onyx*, and *G. acuta*. *G. onyx* and *G. brimeri* exhibit similar shell morphology, and it is suggested that they were endobyssate, semi-infaunal dwellers in soft muds. In these protected low-energy environments they did not need a strong byssus to maintain a stable life position with more than half of the shell buried in the sediment. *G. acuta*, which has highly flattened shell in dorso-ventral direction, is interpreted to have been without byssus. Its flattened shell enabled it to maintain a stable position in a quiet environment. Its conspicuous shell sculpture of fine concentric threads and its additional pattern of radial riblets possibly had a stabilizing function.

Goniophora tyri, which is also a medium-sized species, occurs in in siltstones and coarse-grained limestones, and has a shell form similar to that of *G. onyx* and *G. brimeri*. It is the only modiomorphid to exhibit a conspicuous shell pattern of undulating costellae. This sculpture possibly strengthened the shell, but it may also have been stabilizing in an environment of higher water energy.

'*Goniophora*' *gymeri* and '*G.*' sp. 1 have been found in fore-reef lithologies. They have large shells and are particularly characterized by their strongly reduced anterior lobe, which is almost absent in '*G.*' *gymeri*. The latter species also shows an indented ventral margin. Both species have a relatively high umbo with a strong umbonal shelf for the attachment of the anterior adductor muscle. They also have a highly expanded posterior part of the shell, all features typical of epifaunally attached mytilids of today. They seem to have been well adapted to the high-energy environments of the fore-reef, presumably attached to the reef itself or other hard surfaces.

Fig. 25 shows a reconstruction of the life position of some of the remaining modiomorphids from the Burgsvik Beds of Gotland.

The bivalve fauna at Mulde is highly diverse containing in addition to the four modiomorphids studied, eight nuculoids and about a dozen undescribed species of other bivalve families. *Colpomya hugini* and *Aleodonta burei* occur in great numbers and together with the protobranch nuculoid *Nuculoidea pinguis*, they dominate this community.

The original substrate was a soft carbonate mud mixed with terrigenous clay. The abundance of protobranch bivalves, gastropods, polychaete worms, etc., suggests that it was thixotropic or in a fluid state in its upper part (reworking of modern such sediments by burrowing organisms results in a soupy upper part of the substrate; see Rhoads & Young 1970). In such a physically unstable sediment there is, in addition to the danger of clogging, a significant problem for shell-bearing macrofaunal organisms to avoid sinking down into the toxic zone (Rhoads 1970, p. 461). One way of avoiding sinking is through rapid, effective burrowing (Rhoads 1970; Thayer 1975). Another is to have a more or less elongate shell, which can be embedded with the posterior end above the water–sediment interface (Thayer 1975).

All modiomorphids of the Mulde community show muscular impression patterns typical of burrowing bivalves, and all but *Aleodonta burei* also have a shell shape indicating fairly effective burrowing capacity (Fig. 23:1, 2, 6). They are all interpreted to have been semi-infaunal, byssally attached with their posterior part protruding well above the sediment surface (Figs. 38, 40, and 47).

Stratigraphic distribution

According to the literature, bivalves are often scarce in the Silurian sequences of Gotland (Hede 1921, 1925, 1927a, b, 1940, 1960) and Scania (Grönwall 1897; Moberg & Grönwall 1909; Hede 1915). Reported bivalve diversity, however, may not necessarily reflect the actual diversity in Silurian biota. In the marls of Gotland, for instance, bivalves are seldom encountered, owing to the fact that molluscan shells are prone to early diagenetic dissolution in these sediments. Molluscan fossils recorded in the marls are almost without exceptions preserved as internal moulds (steinkerns). Another consequence of this is that the hinge of bivalves will not be preserved, and taxonomic identification must be based only on overall shape. In such cases the bivalves encountered are almost impossible to refer to any specific taxon, and bivalves are therefore unrepresented in faunal lists and other records.

In the oolitic sediments of the Burgsvik Beds, on the other hand, molluscan and other carbonate structures are well preserved, probably because of oversaturation of calcium carbonate in the original environment. These sediments may show a truer picture of molluscan diversity than the marls.

Bivalves (except for a few epifaunal species) are conspicuously lacking in those stratigraphical units representing reef building periods, i.e. Högklint reefs, Slite reefs (Lower Slite Beds), Lower Hemse Beds, Hamra Beds and Sundre Beds.

Nuculoida

Nuculoids are not common in the Silurian of Gotland and are mostly found in the marls. Generally they constitute an insignificant part of the bivalve faunas, but occasionally they may dominate them totally, in number of individuals as well as in number of species (e.g., the Möllbos fauna).

The comparatively high diversity of nuculoids in the Mulde Beds at Djupvik and at the former Mulde brickyard (Mulde tegelbruk) may be due to intensive collecting during a whole century. These localities are possibly some of the most prolific Silurian fossil localities known.

The conspicuously high diversity of nuculoids recorded in the Halla Beds at Möllbos, probably redflects a true abundance in a fine-grained sediment with a high organic content (Liljedahl 1985).

With a few exceptions, nuculoids, being soft bottom dwellers, are missing from those stratigraphical units representing reef-building events, i.e. Högklint reefs, Slite reefs (Lower Slite Beds), Lower Hemse Beds, Hamra Beds and Sundre Beds (cf. Eriksson & Laufeld 1978, Fig. 16).

The stratigraphical occurrences of the nuculoid species on Gotland are diagrammatically presented in Fig. 26.

Lower Visby Beds. – The lowermost unit of the Gotland sequence exposed on land, the Lower Visby Beds, has so far only yielded one nuculoid species, viz. *Nuculodonta gotlandica.* The specimens, mostly internal moulds, are generally badly preserved. However, one limonitic specimen retains the external shell characteristics (Fig. 28P), and a few internal moulds show distinctive features sufficient for identification (28B).

Upper Visby Beds. – The Upper Visby Beds have yielded two nuculoid species, *Tancrediopsis gotlandica* and *Nuculoidea pinguis,* both having the same stratigraphical distribution reaching up into the Mulde Beds.

Högklint Beds. – In the Högklint Beds only one nuculoid species has been recovered, *Nuculodonta gotlandica.*

Tofta Beds. – No nuculoids have been found in the Tofta Beds. This unit is generally characterized by extremely shallow-water sediments (Hede 1940, p. 43) and lack of macrofossils.

Slite Beds. – The Slite Beds have produced two species of the genus *Tancrediopsis*: *T. gotlandica* and *T. solituda,* and also *Nuculoidea pinguis.* This unit also contains a large size variety of *Nuculoidea* resembling both *N. lens* and *N. pinguis.* As these specimens are preserved only as steinkerns, they cannot with certainty be assigned to any species and are therefore not included in the diagram of Fig. 26.

Halla Beds. – The Halla Beds contain a nuculoid fauna consisting of five species: *Nuculodonta gotlandica, Palaeostraba baltica, Nuculoidea lens, Nuculoidea* sp. A, and *Caesariella lindensis.* Two of these, *Palaeostraba baltica* and *Nuculoidea* sp. A, have only been found in this unit, whereas the occurrence of *Nuculodonta gotlandica* marks the end of its stratigraphical range, and *Caesariella lindensis* and *Nuculoidea lens* make their first appearance at this level.

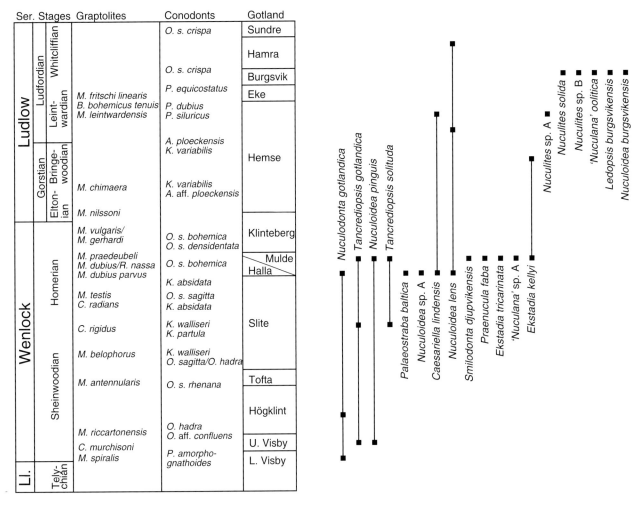

Fig. 26. Stratigraphical distribution of the Silurian Nuculoida of Gotland. Conodont zonation after *L. Jeppsson, V. Viira and P. Männik (unpublished)*.

Mulde Beds. – The nuculoid fauna from the Mulde Beds is by far the richest found on Gotland, containing eight species: *Tancrediopsis gotlandica, T. solituda, Nuculoidea pinguis, Similodonta djupvikensis, Praenucula faba, Ekstadia tricarinata, Ekstadia kellyi* and '*Nuculana*' sp. A. Three of these, *Tancrediopsis gotlandica, Tancrediopsis solituda* and *Nuculoidea pinguis,* have not been found in younger units, while the range of *Ekstadia kellyi* continues up into the Hemse Beds. The remaining forms have only been found in the Mulde Beds.

Klinteberg Beds. – The Klinteberg Beds have not yielded any nuculoid species.

Hemse Beds. – Four nuculoid species have been recorded from the Hemse Beds: *Caesariella lindensis, Nuculoidea lens, Ekstadia kellyi* and *Nuculites* sp. A. Of these only *Nuculites* sp. A has been found in this unit alone, whereas the occurrences of *Ekstadia kellyi* and *Caesariella lindensis* constitute their final appearances in the sequence. The range of *Nuculoidea lens* continues into the Hamra Beds.

Burgsvik Beds. – The next younger unit to contain nuculoids is the Burgsvik Beds, where five species have been recorded: *Nuculites solida, N.* sp. B, '*Nuculana*' *oolitica, Ledopsis burgsvikensis* and *Nuculoidea burgsvikensis*. They were all isolated from coarse sediments such as sandstone, sand, and oolitic limestones.

Hamra and Sundre Beds. – The Hamra and Sundre Beds mainly represent reef-forming events, and no nuculoids have been found so far.

Modiomorphidae

The Modiomorphidae comprise almost $\frac{1}{3}$ of the estimated total number of bivalve species from the Silurian of Sweden (about 100). They occur in 7 of the 13 Gotland units and in the uppermost 2 of the 5 Silurian units in Scania. The oldest occurrence is either in the Llandoverian or in the Wenlockian part of the Visby marl on Gotland, and the youngest is in the Pridolian Öved Sandstone in Scania.

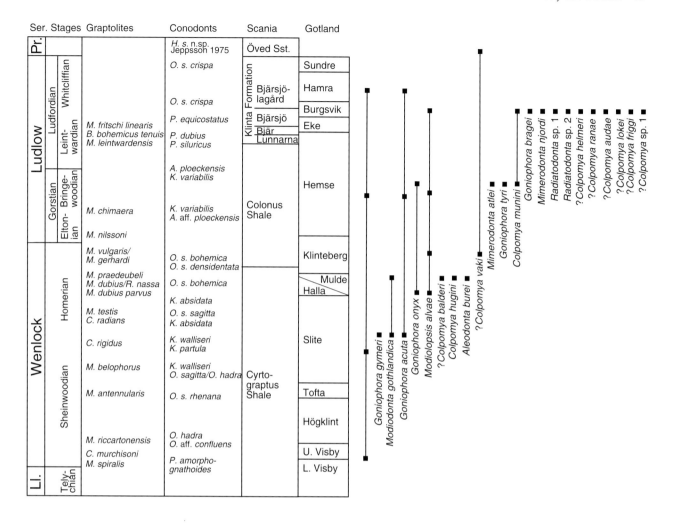

Fig. 27. Stratigraphical distribution of the Silurian Modiomorphidae of Sweden. Conodont zonation after L. Jeppsson, V. Viira and P. Männik (unpublished).

The stratigraphical occurrence of the Modiomorphidae is diagrammatically presented in Fig. 27.

Visby Beds. – The lowermost units of the Gotland sequence, the Lower and Upper Visby Beds, have only yielded one modiomorphid species, viz. *Goniophora brimeri*, which is represented by two articulated internal casts. The exact stratigraphical location of the collection is unknown, and thus it cannot be established if they originate from the Lower or Upper Visby Beds. *G. brimeri* has by far the longest stratigraphical range of the modiomorphids studied, in that it ranges into the Whitcliffian Klinta Formation in Scania.

Högklint Beds. – Pojeta (1979, p. 110, Fig. 33) recorded *Goniophora* sp. a and *Modiomorphidae?* sp. from the Högklint Beds of the Vattenfallet section on Gotland. His specimens are badly preserved and cannot be assigned to any of the species studied, none of which have been found in this unit.

Tofta Beds. – No modiomorphids have been found in this unit. The Tofta Beds are commonly remarkably poor in

macrofossils (Hede 1921, p. 37). The sediments of this unit are indicative of extreme shallow-water conditions (Hede 1940, p. 43).

Slite Beds. – Four modiomorphid species have been recognized in the Slite Beds, viz. *Modiodonta gothlandica, Goniophora brimeri, G. acuta,* and '*Goniophora*' *gymeri*. The first three were isolated from fine-grained, marly sediments that probably represent low energy environments, while '*Goniophora*' *gymeri* was found in a coarse bioclastic limestone, which appears to have been deposited in high-energy environments.

The occurrence of *Modiodonta gothlandica* in the Slite Beds represents its earliest appearance. It ranges into the Mulde Beds. Also *Goniophora acuta* has its first appearance in the Slite Beds, and it ranges as high as the Ludlovian Klinta Formation in Scania. '*Goniophora*' *gymeri* is the only species restricted to the Slite Beds. All these species are represented by a single specimen each in the Slite Beds.

Halla Beds. – *Goniophora onyx* appears for the first time in this unit. In the silicified fauna of Möllbos 1 (Liljedahl 1983, 1984, 1985, 1986), this species is the best preserved of all modiomorphids recorded in Sweden, and accordingly it is the best known morphologically (42 specimens). *Modiolopsis alvae* is the second modiomorphid species known from well preserved silicified specimens and is represented by about a dozen specimens. The range of *G. onyx* reaches up to the Hemse Beds, while that of *M. alvae* extends up to the Burgsvik Beds.

Mulde Beds. – Four modiomorphid species have been reported from the Mulde Beds. The occurrence of *Modiodonta gothlandica* marks the upper end of its stratigraphical range. It is a member of a fauna of four modiomorphid species, three of which make their only appearance in this unit, viz. ?*Colpomya balderi*, *Colpomya hugini*, and *Aleodonta burei*. *Modiodonta gothlandica* is represented by 18 specimens, ?*Colpomya balderi* by 2, and *Colpomya hugini* and *Aleodonta burei* by hundreds of specimens each.

Klinteberg Beds. – The Klinteberg Beds have yielded two modiomorphid species, both of which have long stratigraphic ranges: *Modiolopsis alvae*, which ranges from the Halla Beds to the Burgsvik Beds, and ?*Colpomya vaki*, which ranges up to the Pridolian Öved Sandstone in Scania.

?*Colpomya vaki* is represented by about 20 specimens in the Gotland Klinteberg Beds, and about 30 specimens in the Öved Sandstone at Helvetesgraven in Scania.

Hemse Beds. – In the Hemse Beds seven modiomorphid species have been recorded. The occurrence of *Goniophora onyx*, which is represented by only a single specimen in this unit, constitutes its final appearance in the Gotland sequence, while *Modiolopsis alvae* continues up into the Burgsvik Beds. *Goniophora brimeri* and *Goniophora acuta* reach the Klinta Formation. *Goniophora acuta* is also represented by only a single specimen in the Hemse Beds. *Mimerodonta atlei* (20 specimens) and *Goniophora tyri* (9 specimens) are the only species restricted to the Hemse Beds. The only known specimen of *Colpomya munini* from this unit is questionably referred to this taxon, which in its final appearance in the Burgsvik Beds is represented by about 20 specimens.

Eke Beds. – No modiomorphids have been found in this unit.

Burgsvik Beds. – The Burgsvik Beds contain the highest diversity of modiomorphids in the Gotland sequence, a total of 12 species. For *Modiolopsis alvae* and *Colpomya munini* this is the youngest record, while the rest of the Burgsvik modiomorphids are found exclusively in this unit. These include ?*Colpomya heimeri*, ?*C. ranae*, ?*C. audae*, ?*C. lokei*, ?*C. friggi*, ?*C.* sp. 1. *Mimerodonta njordi*, *Radiatodonta* sp. 1, *R.* sp. 2, and *Goniophora bragei*. The specimens of *Radiatodonta* are by far the largest of all modiomorphids, and the two species of this genus are represented by only a single specimen each. Fig.

25 shows a reconstruction of some of the members of the bivalve faunas of the Burgsvik Beds.

Hamra and Sundre Beds. – The Hamra and Sundre Beds represent reef building periods. In such sediments bivalves are, as a whole, lacking on Gotland. No modiomorphids have been reported from these beds.

Klinta Formation. – In the Klinta Formation of Scania the following modiomorphids have been recorded: *Goniophora brimeri*, which is represented by a few specimens only; *G. acuta* which is known from two articulated specimens, one of which occurs *in situ* in suggested life position (Fig. 54H).

Öved Sandstone. – The Öved Sandstone at Helvetesgraven in Scania contains numerous articulated specimens of ?*Colpomya vaki*. This species ranges from the Klinteberg Beds in the Gotland sequence, and the occurrence in Scania marks the upper end of its range as well as the top of the range of the Modiomorphidae in Sweden.

Classification

Nuculoida

Nuculoid suprageneric classification has been discussed by various authors (e.g., Pelseneer 1891; Dall 1913; Douvillé 1913; Cox 1960; Newell 1969; McAlester 1969).

Classification in this paper follows that of Newell (1969) in general. However, some species do not fit into this classification, i.e. they exhibit morphological characteristics typical of more than one superfamily (see Liljedahl 1983, p. 31). Some have features which do not correspond to the diagnosis of the different higher taxa-groups as they now stand (see, e.g., discussion on *Similodonta* and *Nuculodonta*).

Recently Pojeta (1988) suggested that some genera, previously considered to be nuculoids and (by McAlester 1969) placed in the family Ctenodontidae, are actually solemyids. The remaining genera of the Ctenodontidae, including the type genus *Ctenodonta* Salter, 1852, fit fairly well the diagnosis of the superfamily Nuculanacea (as proposed by McAlester 1969, pp. N231–235). Consequently, the superfamily Ctenodontacea (containing the family Ctenodontidae only) may be superfluous (Pojeta 1988, p. 211).

Modiomorphidae

Bivalves referred to as 'modiomorphids' constitute an aggregate of species with roughly the same external shell morphology. The classification of many such taxa is uncertain, owing to lack of knowledge of the internal features.

The systematic position and evolutionary role of the whole group have long been controversial. Some have suggested that the group is related to carditaceans (Newell 1957; Bailey 1983), to the actinodonts (Douvillé 1913; Newell 1965), to the mytilaceans (Soot-Ryen 1955; Cox 1960; Pojeta 1971; Stanley 1972), or possibly to the crassatellacean (Bailey 1983).

However, Pojeta *et al.* (1986) clarified that the similarity of *Modiomorpha* Hall & Whitfield (type genus of Modiomorphidae) to mytilaceans, and its lack of lateral teeth, indicate that it is not closely related to Carditacea and Crassatellacea. Instead, Pojeta (1971) suggested that early Palaeozoic modiomorphids are ancestral mytilaceans and proposed that these be grouped together in the subclass Isofilibranchia.

Members of the Isofilibranchia, as defined by Pojeta & Gilbert-Tomlinson (1977), are characterized as 'equivalved, inequilateral byssate bivalves with opisthodetic elongate ligament, hinge edentulous or with dysodont teeth, anisomyarian'.

Although some of the modiomorphids studied here differ from the diagnosis given for the subclass (e.g., *Colpomya hugini* and *Aleodonta burei* are slightly inequivalved and have comparatively large and robust hinge teeth, and some species, such as *Goniophora acuta* and ?*Colpomya ranae*, are considered byssate) they are placed in the subclass Isofilibranchia with formal redefinition herein.

Pojeta (1985, p. 105, Fig. 2) considered the stem of Palaeozoic Isofilibranchia to consist of the family Modiomorphidae (extinct in the late Palaeozoic), from which branched off the family Colpomyidae (extinct in the Late Ordovician), the family Orthonotidae (which possibly evolved into the Solenacea), the Recent family Mytilidae, and the superfamily Pholadomyacea (which possibly evolved into the Anomalodesmata).

I propose the new subfamily Modiomorphinae, with the type genus *Modiomorpha* Hall, to contain species with hinge teeth, and the new subfamily Modiolopsinae, with the type genus *Modiolopsis* Hall, to contain species lacking hinge teeth. R.C. Frey (unpublished) suggested the establishment of the Modiolopsinae, to include the edentulous forms of the family. The hypothetical phylogeny of the Modiomorphidae is presented in Fig. 36.

Systematic palaeontology

Class Bivalvia Linnaeus, 1758

Subclass Palaeotaxodonta Korobkov, 1954

Order Nuculoida Dall, 1889

Superfamily Nuculacea Gray, 1824

Diagnosis (McAlester 1969, p. N229). – Truncate posterior extremity, pallial sinus lacking, resilifer present or absent.

Family Praenuculidae McAlester, 1969

Diagnosis (McAlester 1969, p. N229). – External ligament, resilifer absent.

Genus Nuculodonta Liljedahl, 1983

Type and only species. – *Nuculodonta gotlandica* Liljedahl, 1983.

Diagnosis. – Praenuculid with umbones in posterior half of shell, slightly prosogyrate beaks, lunule and escutcheon well defined, conspicuous external ligament posterior to umbo, two series of hinge teeth interrupted by a non-denticulate resilifer-like structure.

Remark. – Although *Nuculodonta gotlandica* exhibits a resilifer-like structure, indicating the presence of an additional internal part of the ligament, this taxon is placed in the family Praenuculidae in spite of the diagnosis given above.

Nuculodonta gotlandica Liljedahl, 1983
Figs. 7, 11:1, 12:1, 15, 26, 28

Synonymy. – □1964 *Ctenodonta sp. A* – Soot-Ryen, p. 47, Pl. 2:1. □1983 *Nuculodonta gotlandica* n.sp. – Liljedahl, pp. 33–47, Figs. 10–18. □1984a *Nuculodonta gotlandica* Liljedahl – Liljedahl, pp. 6–12, Figs. 1–5. □1984b *Nuculodonta gotlandica* Liljedahl – Liljedahl, pp. 54–57, 60–61, 64, Figs. 1–7. □1984c *Nuculodonta gotlandica* Liljedahl – Liljedahl, pp. 2, 7–8, 18–20, Fig. 1. □1986 *Nuculodonta gotlandica* Liljedahl – Liljedahl, Figs. 1, 3, 5.

Holotype. – SGU Type 1056, a left valve, Fig. 28M.

Type stratum. – Halla Beds, Late Wenlockian.

Type locality. – Möllbos 1, Gotland.

Material. – Approximately 300 silicified specimens and 20 badly preserved steinkerns.

Diagnosis. – Nuculoid with the umbones in posterior half of shell, beaks slightly prosogyrate, lunule and escutcheon well defined, conspicuous opisthodetic external ligament; two series of hinge teeth interrupted by a non-denticulate resilifer-like structure.

For a more detailed description, see Liljedahl (1983).

Remarks. – *Nuculodonta gotlandica* has an external ligament on the short side of the valve, thus considered the posterior one. The configuration of the muscular impressions of the foot and its accessory muscles, and also a ridge possibly indicating the position of gill attachment muscles, support the orientation proposed. Moreover, the region of maximum opening of the valves of this species, being in the long part of the valve, agrees with this orientation.

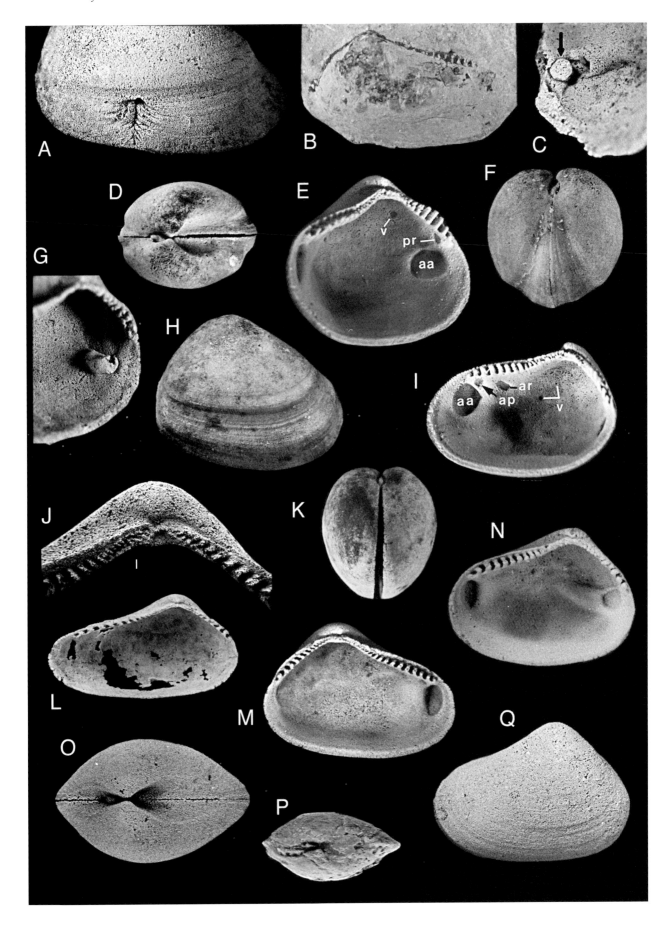

In lateral outline the shell is anteriorly expanded (Fig. 28Q). Shell shape may vary considerably (Fig. 28L, N). The lunule and escutcheon are generally well preserved in the silicified material (Fig. 28O), whereas in moulds of other material they are hardly visible (Fig. 28P). The external opisthodetic ligament is conspicuous (Fig. 28D, F, K, Q), whereas the internal part is delicate and seldom preserved (Fig. 28J).

The anterior hinge plate is longer than the anterior one in adults and carries teeth that are lower and narrower than those of the anterior plate (Fig. 28E, I, M). One of the largest specimens (15.3 mm long) has about 20 teeth in the anterior hinge plate and about 13 in the posterior one (cf. Liljedahl 1983, p. 36, Fig. 12).

In the material of *Nuculodonta gotlandica*, 8 different muscular impressions have been recorded (Fig. 7a). The pertinent soft-part reconstruction is shown in Fig. 7b (see further discussion in 'Morphology of the nuculoid shell').

Most silicified valves show a conspicuous anterior adductor muscle scar (Fig. 28I, N, M)), while the less impressed posterior adductor muscle scar is not commonly preserved (Fig. 28E, M–N)). Traces of accessory muscles are rare, but in some specimens they are evident. The largest of these scars is that of the posterior pedal retractor muscle (No. 3 in Fig. 7a; Fig. 28E) and the second largest that of the anterior pedal retractor muscle (No. 4 in Fig. 7a; Fig. 28I). The impression of the anterior pedal protractor muscle is also usually evident (No. 5 in Fig. 7a; Fig. 28I). Rarely, a ridge extending anteriorly from the posterior adductor muscle scar is present (Fig. 28N). This is assumed to be the site of gill attachment muscles (Fig. 7b).

Regular external growth lines are generally visible (Fig. 28A, H, Q). Periods of slow growth are recorded in some specimens as comarginal bands on the shell sculpture (Fig. 28D, F, H; see also discussion below).

Occurrence. – Late Llandoverian Lower Visby Beds at Norderstrand, Early Wenlockian Upper Visby Beds at Visby, and Late Wenlockian Halla Beds at Möllbos 1, Gotland.

Genus *Ledopsis* Beushausen, 1884

Type species. – *Ledopsis rectangularis* Beushausen, 1884

Ledopsis burgsvikensis (Soot-Ryen, 1964)
Figs. 11:2, 14:1, 15, 26, 29E–G

Original combination. – *Ctenodonta burgsvikensis* Soot-Ryen, 1964, p. 496.

Holotype. – RMMo 150273, Fig. 29E–F.

Type stratum. – Burgsvik oolite, Burgsvik Beds, Late Ludlovian.

Type locality. – Burgsvik, Gotland.

Material. – Four specimens, single valves, shells preserved as recrystallized calcite.

Diagnosis. – Nuculoid shell with expanded posterior end, beaks prosogyrate, evident umbonal ridges; large anterior hinge teeth, smaller posterior ones, tooth row continuous below beak; no resilifer.

Description. – Shell small, subtriangular, moderately inflated, posterior end protruding, tapering, posterior margin forming a conspicuous acute angle; beaks prosogyrate, posterior umbonal ridge reaching to postero-ventral angle; no lunule; no escutcheon; hinge containing anterior and posterior chevron-shaped teeth with apices towards beak, anterior ones large and robust, posterior ones small, continuous below beak, proximal one almost lamellar; no bifurcating teeth; no resilifer; no further internal features observed; external opisthodetic ligament.

Remarks. – *Ledopsis burgsvikensis* has an external ligament on the long side of the valve thus considered to be posterior.

Dimensions of holotype. – Length = 11.5 mm, height = 8.3 mm, width = 3.3 mm (single valve), H/L = 0.72, W/L = 0.29 (both valves = 0.57).

Fig. 28. Nuculodonta gotlandica Liljedahl, 1983. □A. External view showing old shell fracture and also deformed margin caused by damage to mantle which recovered and eventually produced a normal shell margin. SGU Type 1884, ×4.8, sample G78-2LL. □B. Internal mould of right valve showing hinge dentition. RMMo 15777, ×2.9, Lower Visby Beds, Visby. □C. Blister pearl (oldest known bivalve pearl, at arrow) close to adductor muscle scar. SGU Type 3950, ×5.0, sample G79-115Lj. □D. Dorsal view of articulated specimen showing deep comarginal furrows caused by periods of retarded growth, same specimen as F, SGU Type 1194/95, sample G78-1LL. □E. Antero-ventral view exhibiting posterior adductor muscle scar (aa), visceral attachment muscle scar (v), and posterior pedal retractor muscle scar (pr; cf Fig. 7). ×4.0, sample G79-82LJ. □F. Posterior view of same specimen as D, ×4.2. □G. Tumour caused by commensal close to posterior adductor muscle scar, SGU Type 1030, ×5.4, sample G78-2LL. □H. A right valve with evident comarginal growth stops and thin growth lines. SGU Type 999, ×4.3, sample G78-1LL. □I. Postero-ventral view of right valve showing anterior adductor muscle scar (aa), anterior pedal protractor muscle scar (ap), anterior pedal retractor muscle scar (ar), and visceral attachment muscle scars (v; cf. Fig. 7), same specimen as E, ×4.0. □J. Lateral view of right valve showing detail of hinge with resilifer-like structure below the beak, SGU Type 1026, ×13.0, sample G78-2LL. □K. Posterior view of articulated specimen, SGU Type 1060/61, ×4.0, sample G77-28LJ. □L. Lateral view of right valve, SGU Type 1000. ×3.6, sample G78-1LL. □M. Lateral view of holotype (left valve), SGU Type 1056, ×3.9, sample G77-29LJ. □N. Lateral view of right valve showing ridge in posterior part of the shell, possibly being the site of gill suspensory muscles (cf. Fig. 7), SGU Type 1198, ×5.0, sample G77-29LJ. □O. Dorsal view of articulated specimen, anterior to the right, SGU Type 2167/68, sample G79-79LJ. □P. Dorsal view of articulated specimen, anterior to the right, RMMo 15779, ×3.1, Lower Visby Beds at Visby. □Q. Lateral view of left valve, SGU Type 1036, ×4.4, sample G78-2LL. All specimens, except for B and P, are from the Halla Beds of Möllbos 1. All specimens, except for B and P, are silicified; B is an internal mould of marl, and P is preserved as a red-stained iron compound. All are photographs, except for J, which is a scanning electron micrograph.

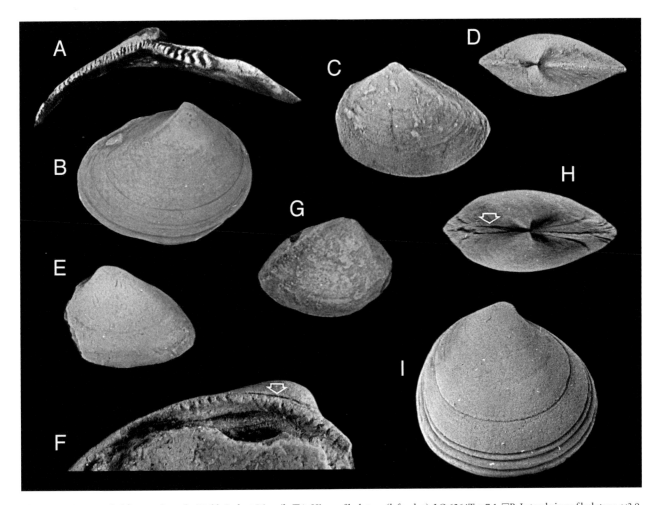

Fig. 29. A–D *Praenucula faba* n.sp. from the Mulde Beds at Djupvik. □A. Hinge of holotype (left valve), LO 6264T, ×7.1. □B. Lateral view of holotype, ×3.9. □C. Lateral view of right valve of articulated specimen, RMMo 21959, ×3.8. □D. Dorsal view of same specimen as C, (anterior to the right), ×3.6. □E–G. *Ledopsis burgsvikensis* (Soot-Ryen 1964) from the Burgsvik Beds at Burgsvik. □E. Lateral view of holotype (left valve), RMMo 150273, ×3.0. □F. Hinge of holotype, ×4.8 (note ligamental groove at arrow). □G. Lateral view of right valve, RMMo 150383, ×2.8. □H–I *Similodonta djupvikensis* Soot-Ryen 1964 from the Mulde Beds at Djupvik. □H. Dorsal view of holotype, RMMo 21934, note ligamental nymphs (at arrow; anterior to the right), ×4.0. □I. Lateral view of holotype, right valve, ×4.2. All specimens preserved as calcium carbonate shells.

Discussion. – In spite of having a truncate anterior extremity (contrary to the diagnosis of Nuculacea), *Ledopsis burgsvikensis* is placed in the Praenuculidae of the Nuculacea, because of its external ligament and continuous series of hinge teeth. Soot-Ryen (1964, p. 497) considered the longer end to be anterior. However, the hinge plate situated in the short end of the shell contains larger teeth than that in the longer end of the shell, indicating that the short end is anterior (See Bradshaw 1971). Furthermore, the furrow along the dorsal margin of the longer end of the shell (Fig. 29F) is interpreted as a groove for the reception of an external ligament, in analogy with living nuculids. Accordingly, this species is suggested to have had truncated anterior extremity, thus fitting the diagnosis of *Ledopsis.*

Occurrence. – Ludlovian Burgsvik Beds at the type locality.

Genus *Similodonta* Soot-Ryen, 1964

Type species. – *Tellinomya similis* Ulrich, 1892.

Discussion. – Although the standard diagnosis of the super-family Nuculacea states 'truncate posterior extremity' (Newell 1969, p. N229), *Similodonta,* in spite of being equilateral, is placed in the family Praenuculidae within the Nuculacea. This is because the hinge consists of two equal series of teeth, continuous below the beak, and lacks resilifer).

Similodonta djupvikensis Soot-Ryen, 1964

Figs. 11:3, 13:2, 15, 26, 29H–I

Holotype. – RMMo 21934, Fig. 29H–I.

Type stratum. – Mulde Beds, Homerian, Wenlockian.

Type locality. – Djupvik, Gotland.

Material. – Only the holotype available at the time of revision. The remaining material described by Soot-Ryen has not been located.

Diagnosis. – Small, subtriangular, height almost equal to length, umbones prominent, beaks small, prosogyrate; no lunule, no escutcheon; long, narrow opisthodetic external ligament; teeth continuous below beak.

For a detailed description, see Soot-Ryen 1964, p. 498.

Remarks. – The beaks point away from the side on which an external ligament is situated. Thus the shell is prosogyrous.

Dimensions of holotype. – Length = 10.6 mm, height = 10.4 mm, width = 5.1 mm (both valves), H/L = 0.98. W/L = 0.48 (both valves).

Occurrence. – Wenlockian Mulde Beds at the type locality.

Genus *Praenucula* Pfab, 1934

Diagnosis. – (Translated from German) General shell shape *Nucula* or *Ctenodonta*-like, integripalliate, below beak between two thick protuberances a furrow, open dorsally and ventrally, interpreted as a possible primitive resilifer; outer ligament always close to internal part; hinge-type III (all teeth chevron-shaped, with their apices towards beak).

Type species. – *Praenucula expansa* Pfab, 1934

Praenucula faba n.sp.

Figs. 11:4, 13:8, 15, 26, 29A–D

Derivation of name. – From Latin *faba* meaning bean. Already in the middle of the last century Lindström sorted out and labelled specimens of this species which he called *Ctenodonta faba*. These were later included in the collections of *Nuculoidea pinguis pinguis* studied by Soot-Ryen (1964).

Holotype. – A left valve, LO 6264, L = 12.2 mm, H = 9.8 mm. H/L = 0.80, Fig. 29A–B.

Type stratum. – Mulde Beds, Homerian, Wenlockian.

Type locality. – Djupvik 1, Gotland.

Diagnosis. – Shell small, subcircular, anterior end expanded, beaks inconspicuous, prosogyrate, distinct anterior umbonal ridge, teeth continuous below beak, posterior teeth more numerous than anterior ones.

Material. – About 50 articulated specimens and one single valve with shells preserved as recrystallized calcite. Nine measured specimens have an average height/length ratio of 0.81, width (both valves)/length ratio of 0.44.

Description. – Shell medium-sized, subcircular, equivalve, inequilateral, compressed, anterior part protruding, margins even, parivincular opisthodetic ligament; shell surface smooth with faint concentric growth lines, sometimes with evident growth increment stops; beaks small, close together, prosogyrate, in posterior half of shell; maximum convexity about midway along shell and slightly above mid height of shell; lunule circumscribed; no escutcheon; conspicuous anterior umbonal keel, anterior margin well rounded, narrow; posterior margin well rounded; ventral margin convex; dorsal margin with conspicuously raised edges; hinge line convex; anterior hinge plate broad, containing in holotype (L = 12.2 mm) approximately 9 chevron-shaped teeth, apices towards beak; posterior hinge plate narrow, in holotype containing about 25 lamellar teeth, the most proximal ones chevron-shaped; hinge plates forming a continuous series below beak; no resilifer; hinge angle approximately 140°; no further internal features observed.

Remarks. – The short side of the valve has an external ligament and a hinge plate with smaller and lower teeth than the other side, which suggests the short side to be posterior.

Comparisons. – *Praenucula faba* has a more circular lateral outline and a less conspicuous umbo than the type species of the genus, *P. expansa* Pfab, 1934, Fig. 3:10–11.

Remark on systematic position. – The hinge of *Praenucula* was classified as hinge type III by Pfab (1934, p. 234), viz. both anterior and posterior hinge teeth are chevron-shaped with their apices towards the beak (Pfab 1934, p. 203). Although most of the posterior teeth of *Praenucula faba* are lamellar, the proximalmost ones are chevron-shaped like those on the anterior hinge plate. Thus its hinge agrees fairly well with that of *Praenucula* given in the diagnosis.

Occurrence. – Wenlockian (Homerian) Mulde Beds at Djupvik, Mulde former brickyard and Gandarve, Gotland. All specimens isolated from marls.

Family Nuculidae Gray, 1824

Diagnosis. – (McAlester 1969). – Shell material nacreous, beaks opisthogyrate, ligament predominantly internal (see Owen 1959), resilifer present.

Genus *Nuculoidea* Williams & Breger, 1916

Type species. – *Cucullea opima* Hall, 1843 (synonymy *Nucula randalli* Hall, 1869).

Remarks. – Williams & Breger (1916, p. 173) proposed *Nuculoidea* as a new subgenus of *Nucula* Lamarck, characterized by the presence of a resilifer and the absence of a pectinate internal margin. They included shells with variable outline and with orthogyrate to slightly prosogyrate umbos (*Nuculoidea opima*) as well as shells with opisthogyrate umbos (*Nucula aquisgranensis* Beushausen).

Vokes (1949) and McAlester (1962) showed, however, that shells of the type species *Nuloidea opima* did exhibit micropectinate internal margins, and concluded that the presence or absence of this feature lacks generic value.

The Silurian *Nuculoidea lens*, examined herein in hundreds of extremely well-preserved silicified and calcitic specimens, exhibits a smooth internal margin. This confirms the suggestion by Vokes (1949) and McAlester (1962) concerning the nature of the internal margin of *Nuculoidea*.

Bailey (1983, p. 247, 1986, p. 1178) and Liljedahl (1983, p. 14) presented each a revised diagnosis of the genus *Nuculoidea*. These are combined here as follows.

Emended diagnosis. – Nuculid with erect to only slightly opisthogyrous or only slightly prosogyrous umbo; outline of shell variable; shell surface smooth or with fasciculate growth lines, sometimes with fine, discontinuous radial microribbing; inner margin either smooth or with micropectinations; microribbing and micropectinations best defined posteriorly; resilifer small, triangular, erect to slightly reclining, sometimes protruding into valve interior and lying beneath the inner ends of the anterior and posterior tooth rows; resilifer showing little to moderate excavation into the hinge plate; hinge plates variable in size and containing a varying number of teeth; anterior plate containing more and larger teeth than posterior plate.

Nuculoidea lens Liljedahl, 1983

Figs. 6, 11:5, 12:2, 15, 26, 30

Synomymy. – ☐*non* 1880 *Ctenodonta pinguis* Lindström – Lindström *in* Angelin & Lindström, p. 19, Pl. 19:15–16. ☐*non* 1921 *Nucula anglica* d'Orbigny – Hede, p. 32. ☐*non* 1927a *Nucula anglica* d'Orbigny – Hede *in* Munthe *et al.*, pp. 15, 31, 52. ☐1927b *Nucula anglica* d'Orbigny – Hede *in* Munthe *et al.*, pp. 20, 21, 25, 54. ☐*non* 1940 *Nucula anglica* d'Orbigny – Hede *in* Lundquist et. al., pp. 60, 66. ☐1983 *Nuculoidea lens* n.sp. – Liljedahl, pp. 5–31, Figs. 2–9. ☐1984a *Nuculoidea lens* Liljedahl – Liljedahl, pp. 6–11, Figs. 1–5. ☐1984b *Nuculoidea lens* Liljedahl – Liljedahl, pp. 54–55, 60–64. ☐1984c *Nuculoidea lens* Liljedahl – Liljedahl, pp. 2, 7, 18, 20–21. ☐1986 *Nuculoidea lens* Liljedahl – Liljedahl, Fig. 1–2.

Holotype. – SGU Type 842, a single right valve, Fig. 30A, F, G.

Type stratum. – Halla Beds, Upper Wenlockian.

Type locality. – Möllbos 1, Gotland.

Material. – Several hundred silicified specimens.

Diagnosis. – *Nuculoidea*, lens-shaped in dorsal or ventral view and sub-circular to suboval in lateral outline, umbo low,

beaks contiguous, orthogyrate; lunule and escutcheon circumscribed; dorsal margins raised only slightly on both sides of umbo; ventral margin strongly convex; chevron-shaped teeth with apices pointing towards umbo, both anterior and posterior to resilifer; in adults anterior teth more numerous than posterior ones.

For a more detailed description, see Liljedahl (1983, 1984a).

Dimensions of holotype. – Length = 14.8 mm, height = 11.5 mm, width = 4.0 mm, H/L = 0.78, W/L = 0.27 (both valves = 0.54).

Remarks. – *Nuculoidea lens* shows an elaborate pattern of muscular imprints of the foot and its accessory muscles, all indicating that the long end of the valve is anterior (where the point of maximum opening is situated and the foot consequently protruded).

The material of this species is exclusively silicified, and because of the large amount of specimens it has been possible to record a number of morphological characteristics and variations of morphology. Thus, variation of shell shape in lateral outline varies considerably (Fig. 30A, K, M). Lunule and escutcheon may be more or less discernible (Fig. 30E, H).

The hinge teeth are comparatively robust and long (Fig. 30I) and more numerous in the anterior hinge plate than in the posterior one (Fig. 30A, M; cf. Liljedahl 1983, pp. 18–22, Fig. 7). The anterior teeth are longer than the posterior ones, owing to the fact that the point of maximum opening of the valves is in the anterior half of the valve (Fig. 30H).

Impressions of muscles are fairly often well preserved. Fig. 6 shows the maximum number of muscle scars of this species. The anterior adductor muscle scar is the largest and most heavily incised (Fig. 30A, F, K), while the posterior adductor

---

Fig. 30. Nuculoidea lens Liljedahl, 1983. ☐A. Lateral view of holotype (right valve), SGU Type 842, ×3.3, sample G77-28LJ. ☐B. Dorsal view (anterior to the left), note lunule and escutcheon, SGU Types 900/901 (same as C), ×3.9, sample G77-28LJ. ☐C. Lateral view of left valve showing growth increment stops, SGU Type 901, ×3.9. ☐D. Antero-lateral view of left valve showing the posterior pedal retractor muscle scar (at arrows), SGU Type 865, ×5.0, sample G77-28LJ. ☐E. Dorsal view of articulated specimen (anterior to the left), SGU Types 2049/2050, ×4.3, sample G7978LJ. ☐F. Postero-ventral view of holotype showing from left anterior adductor muscle scar (aa), anterior protractor muscle scar (ap), visceral muscle scar (v), anterior pedal retractor muscle scar (ar), and median muscle scar (m), ×3.5. ☐G. Internal antero-ventral view of holotype showing posterior pedal retractor muscle scar (at arrows) and, below it, posterior adductor muscle scar with growth stops, ×3.5. ☐H. Dorsal view of articulated specimens showing maximum opening anteriorly, anterior to the left, SGU Types 1188/1189, ×4.0, sample G78-1LL. ☐I. Ventro-lateral view of right valve with complete dentition in anterior hinge plate, SGU Type 1193, ×4.0, sample G79-78LJ. ☐J. Ventro-lateral view of right valve showing median muscle scars (m), SGU Type 844, ×4.0, sample G77-28LJ. ☐K. Lateral view of left valve, SGU Type 876, ×8.3, sample G77-28Lj. ☐L. Anterior view of SGU Types 814/815, ×4.0, sample G77-28LJ. ☐M. Lateral view of left valve, SGU Type 1137, ×3.5, sample G79-84ALJ. All specimens are silicified and come from the Halla Beds at Möllbos 1.

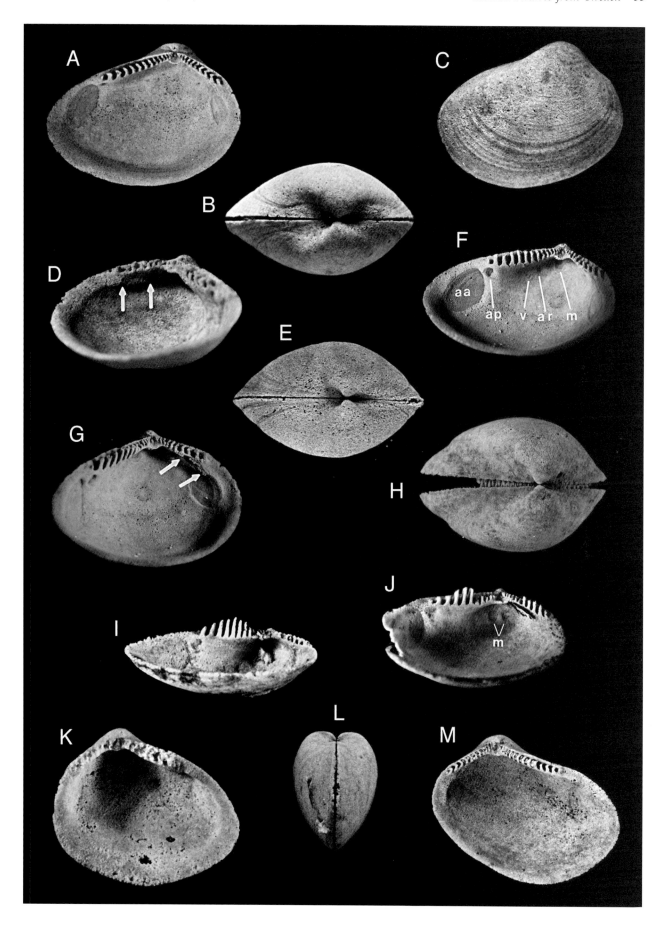

muscle scar is second in size and less deeply impressed than the anterior one. In both scars, growth stops are sometimes present (Fig. 30F–G). Third in size and faintly defined is the posterior pedal retractor muscle scar (Fig. 30D, G). The median muscle scars come next in size but are rarely preserved (Fig. 30F, J). Also rarely encountered are the remaining accessory muscle scars, viz. the anterior pedal protractor muscle scar (Fig. 30F) and the anterior pedal retractor muscle scar (Fig. 30F). The impressions denoted 'v' in Figs. 6 and 30 are assumed to be the site of visceral attachment muscles (Fig. 30F, J).

The species is well known through ontogentic growth series (Liljedahl 1983, 1984a) and morphological and ecological studies (Liljedahl 1984b, 1985).

Comparisons. – For comparisons with closely related species, see Liljedahl (1983, pp. 28–31).

Occurrence. – Wenlockian (Homerian) Halla Beds at Möllbos 1 and at Gothemshammar 7, Ludlovian (Whitcliffian) Hamra Beds at Bottarve 1, Gotland.

Nuculoidea pinguis (Lindström, 1880)

Figs. 11:6, 13:5, 15, 26, 31A–B, D–E, I–J

Synonymy. – ☐1880 *Ctenodonta pinguis* Lindström – Lindström *in* Angelin & Lindström, p. 19, Pl. 19:15, 16. ☐1888 *Ctenodonta pinguis* Lindström – Lindström, p. 12. ☐1921 *Nucula anglica* d'Orbigny – Hede, pp. 32, 41, 48, 51. ☐1927a *Nucula anglica* d'Orbigny – Hede, pp. 15, 31, 52. ☐?1927b *Nucula anglica* d'Orbigny – Hede, pp. 20, 21, 25, 54. ☐1940 *Nucula anglica* d'Orbigny – Hede, pp. 60, 66. ☐1960 *Nucula anglica* d'Orbigny – Hede, pp. 60, 66. ☐1964 *Nuculoidea pinguis pinguis* (pars) (Lindström) – Soot-Ryen, p. 510. ☐*non* 1964 '*Nuculana*' sp. B Soot-Ryen – Soot-Ryen, p. 514. (Soot-Ryen 1964 considered *Nuculoidea pinguis pinguis* and *N. pinguis burgsvikensis* to be varieties of *N. pinguis*. They differ morphologically from each other, however, in several important respects and are therefore considered species in their own right; cf. Liljedahl 1983, pp. 28–31.)

Lectotype. – RMMo 150410, a complete articulated specimen, Fig. 31A, E.

Type stratum. – Mulde Beds, Homerian, Wenlockian.

Type locality. – Djupvik 1, Gotland.

Material. – Several hundred specimens with shells preserved as recrystallized calcite and steinkerns of marl.

Diagnosis. – Shell small, with expanded anterior end, umbones prominent, beaks prosogyrate, hinge with two series of teeth interrupted by triangular resilifer; anterior teeth larger and more numerous than posterior ones.

Description. – For a detailed description, see Soot-Ryen (1964). Because of the abundant material, variation in shell

morphology has been recorded (Fig. 31A, D, J). Of the muscular impressions, that of the anterior adductor is the largest and most deeply incised (Fig. 31J), and that of the posterior adductor comes second in this respect. Also preserved are the pedal protractor muscle scar (Fig. 31B–C), posterior pedal retractor muscle scar (Fig. 31B–C), and some accessory muscle scars in the umbonal cavity (Fig. 31J).

Dimensions of holotype. – Length = 12.6 mm, height = 9.2 mm, width = 6.6 mm (both valves), H/L = 0.73, W/L = 0.52 (both valves)

Remark. – *Nuculoidea pinguis* shows an elaborate pattern of muscular imprints of the foot and its muscles, all indicating that the long end of the valve is anterior (where the point of maximum opening is situated and the foot consequently protruded).

Hede (1921, 1927a, b, 1940) indiscriminately assigned all(?) nuculoids except *Tancrediopsis* to *Nucula* (*Nuculoidea*) *anglica*. Since he did not collect specimens from all localities from which he made faunal lists, it is virtually impossible to check his identifications in all cases. However, in his 1921 paper he reported *Nucula anglica* from the Upper Visby Beds, Slite Beds and Mulde Beds from which only *Nuculoidea pinguis* has been identified among the nuculoids. In Hede 1927 *Nucula anglica* was also reported from the Slite Beds which may also be correct. In Hede's paper from 1927b *Nucula anglica* was reported from the Mulde Beds, which is correct, but also from the Hemse Beds from which strata, however, no collected specimens of *Nucloidea pinguis* have been identified. Possibly *Nucula anglica* is synonymous to

Fig. 31. ☐A. *Nuculoidea pinguis* (Lindström, 1880). Lateral right view of articulated specimen (holotype), RMMo 150410, ×3.1, Mulde Beds at Djupvik. ☐B. *Nuculoidea pinguis*. Dorsal view of steinkern showing dentition and muscular impressions (same lettering as in Fig. 4; anterior to the right), RMMo 15841, ×3.7, Mulde Beds at Djupvik. ☐C. *Nuculoidea* sp. Dorsal view of steinkern showing muscular impressions (same lettering as in Fig. 4), RMMo 15362, ×2.4, Slite Beds at Österby strand. ☐D. *Nuculoidea pinguis* (Lindström, 1880). Lateral right view of articulated specimen, RMMo 15842, ×2.9, Upper Visby Beds at Gnisvärd. ☐E. *Nuculoidea pinguis* Lindström, 1880). Dorsal view of lectotype (anterior to the right), ×3.2. ☐F. *Nuculoidea* sp. Right view of same specimen as in C, ×2.4. ☐G. *Nuculoidea* sp. Right view of steinkern, RMMo 15806, ×2.4, Högklint Beds at Medebys. ☐H. *Nuculoidea* sp. Left view of steinkern, RMMo 15908, ×2.1, Mulde beds at Djupvik. ☐I. *Nuculoidea pinguis* (Lindström, 1880). Hinge of left valve. RMMo 150274, ×6.1, Mulde Beds at Djupvik. ☐J. *Nuculoidea pinguis* (Lindström, 1880). Right view of steinkern showing muscular impressions (same lettering as in Fig. 4), RMMo 15941, ×3.2, Mulde Beds at Djupvik. ☐K–M. *Nuculoidea burgsvikensis* (Soot-Ryen, 1964). ☐K. Hinge of holotype (left valve), RMMo 150384, ×3.3, Burgsvik Beds, probably at Burgsvik. ☐L. Lateral view of right valve showing radial striae, RMMo 150348, ×2.7, Burgsvik Beds at Grötlingbo. ☐M. Lateral view of right valve, RMMo 21758, ×2.0, Burgsvik Beds at Burgsvik. ☐N–O. *Nuculoidea* sp. A. ☐N. Dorsal view (anterior to the left), SGU Types 3836/3837, ×11.7, Halla Beds of Möllbos 1, sample G77-28LJ. ☐O. Lateral left view of same specimen as N, ×11.5. ☐P–Q. ?*Nuculoidea* sp. ☐P. Hinge of left valve, posterior hinge plate damaged, RMMo 15855, ×10.0, locality unknown. ☐Q. External view of same specimen as in No. 16, ×7.1.

Nuculoidea lens in the Hemse Beds. In his paper of 1940 Hede reported *Nucula anglica* from the Slite Beds, which is also probably correct.

Much of the steinkern material from other localities than Djupvik was regarded by Soot-Ryen as belonging to *Nuculoidea pinguis*. At closer scrutiny many of these specimens are impossible, at present, to refer with certainty either to *N. pingius*, *N. lens* or to an unknown species of *Nuculoidea*. This is true especially for those specimens collected from levels where no specimens with a preserved shell have been found and accordingly no certain identification has been possible (Fig. 31C, F–H).

Nuculoidea pinguis differs from *N. burgsvikensis* and *N. lens* as described by Liljedahl (1983, pp. 28–31).

A large size variety, or varieties, of *Nuculoidea*, only preserved as steinkerns, resembles *Nuculoidea pinguis* and *Nuculoidea lens* in gross morphology (Fig. 31C, F–H). The largest specimens examined is 19.9 mm long and the shortest 13.6 mm, while corresponding values of *Nuculoidea lens* are 15.0 mm and 0.5 mm. Maximum length of measured specimens of *Nuculoidea pinguis* is 17.6 mm and smallest 13.6 mm. The H/L ratio of 12 specimens is 0.70–0.80 and the 2W/L relation 0.40–0.48. The corresponding values of *Nuculoidea pinguis* are 0.70–0.80 and steinkerns 0.45, and of *Nuculoidea lens* 0.72–0.91 and 0.60, respectively. This large size variety, or varieties, was found in the Mulde Beds at Djupvik and in the Slite Beds at Österby strand and Storugns kanal.

The two specimens described by Soot-Ryen (1964, p. 514) as '*Nuculana*' sp. B, one of which is illustrated on her Pl. 4:3, show the hinge characteristics of *Nuculoidea*. They probably belong to *Nuculoidea pinguis*, since the relation of the length of the anterior hinge plate to total length of the shell is 0.35, and the length of the posterior hinge plate to total length of shell is 0.2. The corresponding values of *Nuculoidea pinguis* are approximately 0.32 and 0.27. For comparison, those of *Nuculoidea lens* are 0.38 and 0.31, respectively.

Occurrence. – Wenlockian (Homerian) Mulde Beds at Djupvik, Gotland.

Nuculoidea burgsvikensis (Soot-Ryen, 1964)

Figs. 11:7, 14:4, 15, 26, 31K–M

Synonymy. – □1964 *Nuculoidea pinguis burgsvikensis* n. subsp. – Soot-Ryen, p. 512, Pls. 2:7, 4:7, Fig. 6b. (See 'Remark' on *Nuculoidea pinguis*.)

Holotype. – RMMo 15034, a single valve preserved with shell in recrystallized calcite, Fig. 31K.

Type stratum. – Ludlovian (Whitcliffian) Burgsvik Sandstone and Oolite, Burgsvik Beds.

Type locality. – Burgsvik, Gotland.

Material. – Nine specimens, all single valves with shell preserved as recrystallized calcite.

Diagnosis. – Shell large, umbones prominent, beaks orthogyrate to slightly prosogyrate, hinge plates of equal length containing approximately the same number of teeth.

For detailed description see Soot-Ryen 1964, p. 512.

Comparisons. – The comparatively high umbo is a salient feature of *Nuculoidea burgsvikensis* (Fig. 31K–M). The radial striae on the shell surface (Fig. 31L) are rarely preserved, and *Nuculoidea burgsvikensis* is the only species among the nuculoids of Gotland on which such sculpture has been observed. Internally this species is characterized by the two broad, robust hinge plates of equal length (Fig. 31K).

Dimensions of holotype. – Length = 14.3 mm, height = 10.4 mm, width = 4.4 mm, H/L = 0.73, 2W/L = 0.62.

Remark. – For comparisons with closely related species, see Liljedahl (1983, pp. 28–31).

Occurrence. – Ludlovian Burgsvik Beds, Burgsvik, Gotland.

Nuculoidea sp. A

Figs. 11:8, 12:3, 15, 26, 31N–O

One single small, articulated silicified specimen, SGU Type 3836/3837, has been recorded in the Late Wenlockian Halla Beds at Möllbos. It is similar in shell shape to other species of the genus *Nuculoidea* but has, in contrast to these, a conspicuous shell sculpture of concentric costellae (for further discussion, see Liljedahl 1984a, p. 78).

?*Nuculoidea ecaudata* (Soot-Ryen, 1964)

Figs. 11:9, 14:6, 15, 26, 32

Synonymy. – □1964 '*Nuculana*' *ecaudata* n.sp. – Soot-Ryen, p. 513.

Holotype. – RMMo 150275, a single right valve with shell preserved as recrystallized calcite, Fig. 32A–B.

Type stratum. – Gotland, horizon unknown.

Type locality. – Gotland, locality unknown.

Material. – Only the holotype.

Diagnosis. – Shell large, anterior end expanded, umbo low, beaks prosogyrate, anterior hinge teeth broader and more robust than posterior ones, tooth rows interrupted by resilifer.

For detailed description, see Soot-Ryen (1964, p. 513).

Remarks. – ?*Nuculoidea ecaudata*, considered by Soot-Ryen (1964, p. 513) to have had an enlarged posterior end and thus classified as '*Nuculana*' *ecaudata*, is suggested to have the

Fig. 32. ?*Nuculoidea ecaudata* (Soot-Ryen, 1964). □A. Hinge of holotype (right valve), RMMo 150275, ×4,4, locality unknown. □B. Lateral view of holotype, ×2.2.

anterior end enlarged. This is based on the size difference of hinge teeth, where the larger ones are considered to be anterior (cf. Bradshaw & Bradshaw 1971). The umbo of ?*N. ecaudata* is inconspicuous and low, in contrast to the typical large umbo of the genus *Nuculoidea*. However, owing to its hinge and ligament construction, (the larger teeth considered to be anterior; see Bradshaw 1971), it is considered a member of the family Nuculidae and provisionally assigned to *Nuculoidea*.

Dimensions of holotype. – Length = 23.4 mm (not complete), height = 16.3 mm, width = 5.9 mm, W/L = 0.25 (both valves = 0.50).

Occurrence. – The specimen was found in marl, locality on Gotland unknown.

?*Nuculoidea* sp.
Fig. 31P–Q

One single left valve with preserved anterior hinge plate and teeth possibly belongs to *Nuculoidea*. The shell is externally characterized above all by a high and acute umbo and a conspicuous anterior umbonal ridge.

Superfamily Nuculanacea Adams & Adams, 1858

Diagnosis. – (McAlester 1969). – Shell elongate posteriorly, with or without resilifer, pallial sinus usually present.

Family Malletiidae Adams & Adams, 1858

Diagnosis. – (McAlester 1969). – Ligament predominantly external, resilifer lacking.

Genus *Caesariella* Liljedahl, 1984

Type and only species. – *Caesariella lindensis* (Soot-Ryen, 1964).

Diagnosis. – Malletiid with prosogyrate anteriorly extended and posteriorly truncated shell, external ligament posterior to beaks; margins even; teeth forming an uninterrupted series below beak, anterior and posterior part of hinge almost equal, most distal teeth chevron-shaped, apices towards beak, proximal teeth lamellar; pallial line slightly sinuate.

Systematic position. – *Caesariella* shows morphological characteristics typical both for the superfamily Nuculacea and for the superfamily Nuculanacea. Shells belonging to Nuculacea have a truncate posterior extremity and lack pallial sinus, while shells assigned to Nuculanacea have an expanded posterior end and usually also a pallial sinus. Although exhibiting a slightly truncate posterior end, the pallial sinus of *Caesariella* (indicating the presence of siphons) suggests a nuculanacean affinity. Owing to the absence of a true resilifer, it is placed in the Malletiidae.

Caesariella lindensis (Soot-Ryen, 1964)
Figs. 8, 11:10, 12:4, 15, 26, 33

Synonymy. – □1964 *Ctenodontacea lindensis* n.sp. – Soot-Ryen, p. 495, Pl. 1:1–3, 5. □1984a *Caesariella lindensis* (Soot-Ryen) – Liljedahl, p. 20. Figs. 10, 11, 12. □1984b *Caesariella lindensis* (Soot-Ryen) – Liljedahl, pp. 63–64, Fig. 6–7. □1984c *Caesariella lindensis* (Soot-Ryen) – Liljedahl, pp. 2, 9–10, Fig. 1.

Holotype. – RMMo 150356, a right valve, Fig. 33E.

Type stratum. – Hemse Beds, Early Ludlovian.

Type locality. – Linde, Gotland.

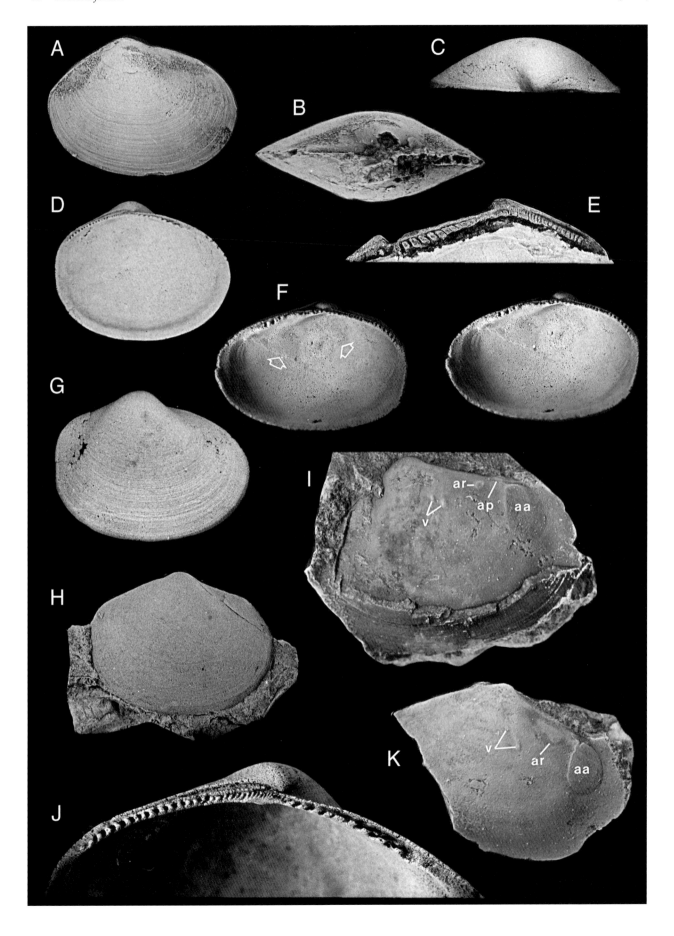

Material. – Approximately 50 specimens, of which 12 are silicified and the remaining ones preserved as recrystallized shells of calcite and internal moulds of fine-grained limestone.

Diagnosis. – Shell large, anterior end expanded, external opisthodetic ligament, umbones low, beaks prosogyrate, escutcheon defined, posterior hinge plate longer and containing more numerous teeth than anterior one, tooth series continuous below beak, edentulous region below and immediately posterior to beak, shallow pallial sinus.

For detailed description see Liljedahl 1984a, pp. 20–26.

Remarks. – *Caesariella lindensis* has, besides an external ligament on the short side of the valve and a pedal muscle pattern similar to that of Recent nuculoids, a conspicuously limited area in the umbonal cavity, reaching the margin in the long part of the valve. It probably represents the extension of the visceral floor, which in living nuculoids coincides with the ventral border of the gonads and encloses the pericardial region with heart, intestines, etc. This region is situated in the anterior part of the valve in Recent nuculoids and thus supports the orientation suggested for *Caesariella lindensis*.

The material available reflects some variation in size and lateral outline of the shell (Fig. 33A, G–H). The Ludlovian specimens from Linde, Mannegårde and Tänglingshällar possibly represent a variety slightly different from the Wenlockian ones from Möllbos (cf. Life habits; Fig. 11:10a,10b). The largest specimen of the Ludlovian variety is 30.9 mm and its shell is less anteriorly expanded (Fig. 33A), has a lower umbo and comparatively fewer posterior teeth in relation to anterior teeth (Fig. 33E) than the largest Wenlockian one (Fig. 18G, J). Maximum length registered among Wenlockian specimens is 15.0 mm.

The umbones are orthogyrate (Fig. 33C), erroneously defined as opisthogyrate by Soot-Ryen (1964, p. 455), and the beaks prosogyrate (Fig. 33C–D, J).

Fig. 33. Caesariella lindensis (Soot-Ryen, 1964). □A. Lateral right view of articulated specimen, RMMo 24152, ×1.6, locality unknown, probably Hemse Beds. □B. Dorsal view of same specimen as A (anterior to the left), ×1.9. □C. Dorsal view of right valve, SGU Type 3606, ×3.4, Halla Beds at Möllbos 1, sample G79-78LJ. □D. Lateral view of left valve, SGU Type 3607, ×3.4, Halla Beds at Möllbos 1, sample G79-78LJ. □E. Hinge of holotype (right valve), RMMo 150356, ×3.2, Hemse Beds at Linde klint. □F. Stereo pair, ventro-lateral view of right valve showing mucular impressions and region of visceral mass (at arows; cf. Fig. 19a), SGU Type 3642, ×3.0, Halla Beds at Möllbos 1, sample G79-79LJ. □G. Lateral view of right valve (same specimen as in C). □H. Lateral view of right valve, RMMo 150352, locality unknown, probably Hemse Beds. □I. Internal mould of right valve showing, from left, two visceral muscle scars (v), anterior pedal retractor muscle scar (ar), anterior pedal protractor muscle scar (ap), and anterior adductor muscle scar (aa), cf. Fig. 19a, RMMo 150330, ×4.4, Hemse Beds at Mannagårda. □J. Hinge of left valve, note ligament nymph posterior to beak, same specimen as D, ×12.0. □K. Internal mould of right valve showing muscular impressions (same lettering as in Fig. 4; cf. Fig. 8), RMMo 24609, ×3,9, Hemse Beds at Tänglingshällar. C, D, F, G, and J are silicified and the remaining specimens preserved as calcium carbonate.

The anterior hinge teeth are longer and more robust than the posterior ones (Fig. 33E, J), both series of teeth continuing into each other. The proximalmost posterior teeth are conspicuously narrow, and below these is an edentulous (resilifer-like?) region (Fig. 33J), which is broader in the left valve.

Several more or less distinct muscle scars have been observed (Fig. 8A), all of which are faintly impressed when compared to nuculoids in general. Largest and deepest is the anterior adductor muscle scar (No. 1 in Fig. 8A; Fig. 33F, I, K). Second in size is the scar of the posterior adductor adductor muscle, which, however, is the faintest defined of all (No. 2 in Fig. 8A; Fig. 33D). Also faintly impressed is the posterior pedal retractor muscle scar (No. 3 in Fig. 8A). The anterior pedal protractor muscle scar is fourth in size (No. 4 in Fig. 8A; Fig. 33D, I), and the anterior pedal retractor muscle scar (No. 5 in Fig. 8A; Fig. 33I, K) is fifth, both being well impressed. Scars Nos. 6 and 7 in Fig. 8A are suggested to be the impressions of visceral attachment muscles. There is a conspicuous ridge enclosing the upper part of the umbonal cavity (Fig. 33F), probably indicating the extension of the visceral floor (Fig. 8B; cf. corresponding feature in modern nuculoids, in Bradshaw 1978). The pallial line has a shallow sinus indicating the presence of only short siphons. Fig. 8B shows the reconstruction of the foot, gills and gonads of *Caesariella lindensis* (cf. Heath 1937, Bradshaw 1978).

Occurrence. – Wenlockian (Homerian) Halla Beds at Möllbos 1 and Ludlovian Hemse Beds at Linde, Mannegårde and Tänglingshällar, Gotland.

Genus *Ekstadia* Soot-Ryen, 1964

Type species. – *Ekstadia tricarinata* Soot-Ryen, 1964.

Species. – *Ekstadia tricarinata* Soot-Ryen, 1964, *E. kellyi* n.sp.

Emended diagnosis. – Shell small to medium-sized, inequilateral, posterior end slightly protruding, umbones prosocline, beaks small, prosogyrate, close together, more or less posterior conspicuous diagonal umbonal sulcus extending to postero-ventral angle, anterior sulcus present or wanting; shell surface smooth with concentric growth lines; lunule and escutcheon not defined; external opisthodetic ligament; hinge teeth in an uninterrupted series, anterior teeth large, broad, robust, more or less chevron-shaped with apices towards beak, posterior teeth small, narrow, chevron-shaped with apices towards beak, posterior teeth more numerous than anterior teeth; fairly deep pallial sinus.

Systematic position. – Soot-Ryen (1964, p. 501) placed *Ekstadia tricarinata* in the Ctenodontidae, most probably because of the external ligament and the character of the hinge (continuous below the beak). The material originally examined by her contains a few specimens of steinkerns showing a comparatively large pallial sinus indicating the presence of siphons. Thus, *Ekstadia* is placed in the Malletiidae of the Nuculanacea.

Ekstadia tricarinata Soot-Ryen, 1964

Fig. 11:11, 13:4, 15, 26, 34A–E

Synonymy. – □1964 *Ekstadia tricarinata* n.sp. – Soot-Ryen, p. 502 (pars).

Holotype. – RMMo 15618, a complete, articulated specimen (Soot-Ryen 1964, Pl.3:3).

Type stratum. – Mulde Beds, Homerian, Late Wenlockian.

Type locality. – Djupvik 1, Gotland.

Material. – About a hundred specimens preserved as shells of recrystallized calcite or steinkerns of marl.

Emended diagnosis. – Shell small with two posterior and two anterior sulci and deep pallial sinus.

New description. – Shell small, quadrangular, two distinct keels from umbo to postero-ventral angle and posterior margin forming a sulcus, two faint keels from umbo to antero-ventral angle and anterior margin, respectively, forming a concave faint sulcus (Fig. 34A–B); umbo dominant, prosocline, beaks small, close together, prosogyrate (Fig. 34C), slightly raised above dorsal margin; shell surface smooth with concentric growth lines; no lunule or escutcheon; dorsal margin convex; anterior margin convex, ventral part truncated; posterior margin rounded, truncated between keels; ventral margin almost straight in anterior part, angular in posterior part; hinge consisting of an uninterrupted series of teeth, anterior hinge plate short, broad with Z-shaped and chevron-shaped teeth with their apices towards beak, a specimen 9.4 mm long containing 4–5 teeth; posterior hinge plate long, narrow with chevron-shaped teeth with apices towards beak, a specimen 9.4 mm long containing approximately 20 teeth (Fig. 34D); impressions of the adductor muscles observed (Fig. 34E), relatively deep pallial sinus (Fig. 34E).

Remarks. – *Ekstadia tricarinata* has a more or less evident external diagonal sulcus in one end also containing a pallial sinus, indicating the presence of siphons. This end is considered to be posterior.

Dimensions of holotype. – Length = 8.0 mm, height = 7.2 mm, 2×width = 4.3 mm, H/L = 0.9, 2W/L = 0.54

Comparisons. – *Ekstadia tricarinata* differs from the new species *E. kellyi* in being smaller, exhibiting a less conspicuous umbo and a pronounced posterior sulcus, having an additional anterior sulcus, and having more numerous hinge teeth in the posterior hinge plate.

Occurrence. – Wenlockian (Homerian) Mulde Beds at Djupvik, Gotland.

Ekstadia kellyi n.sp.

Figs. 10, 11:12, 13:6, 34F–K

Synonymy. – □1964 *Nuculoidea sp. A* – Soot-Ryen, p. 512, Pl. 4:8. □*pars* 1964 *Ekstadia tricarinata* n.sp. – Soot-Ryen, p. 502.

Derivation of the name. – After Kelly, a dear friend.

Holotype. – RMMo 15618, a complete articulated specimen. Length = 12.2 mm, height = 10.4 mm, 2×width = 6.8 mm, H/L = 0.85, 2 W/L = 0.55, Fig. 34H, J.

Type stratum. – Mulde Beds, Homerian, Wenlockian.

Type locality. – Djupvik 1, Gotland.

Material. – Five specimens preserved as recrystallized calcite or steinkerns of marl.

Diagnosis. – Shell medium-sized, quadrangularly subovate, umbo prominent, faint posterior sulcus.

Description. – Shell medium-sized, quadrangular–subovate (Fig. 34H), inequilateral, strongly inflated, umbones prominent, prosocline, in anterior half of shell; beaks small, prosogyrate, close together (Fig. 34J), slightly raised above dorsal margin (Fig. 34K); shell surface smooth with concentric growth lines; dorsal margin slightly convex; anterior margin evenly rounded; posterior margin truncated, straight to slightly convex; ventral margin convex; no lunule or escutcheon; hinge consisting of an uninterrupted series of teeth, anterior hinge plate short, broad with lamellar, slightly chevron-shaped teeth with apices towards beak, a specimen

---

Fig. 34. □A–E. *Ekstadia tricarinata* (Soot-Ryen, 1964) from the Mulde Beds at Djupvik. □A. Lateral right view of articulated specimen, note posterior sulcus (at arrows), RMMo 15510, ×4.6. □B. Lateral left view of articulated specimen, RMMo 15509, ×3.9, note anterior and posterior sulci, respectively (at arrows). □C. Dorsal view of same specimen as in B, note ligament nymph (at arrow), anterior to the left, ×3.8. □D. Hinge of right valve, RMMo 158381, ×6.6. □E. Left view of steinkern showing muscular impressions (same lettering as in Fig. 4, cf. Fig. 10), note deep pallial sinus (at arrow), RMMo 15509, ×5.1. □F–K. *Ekstadia kellyi* n.sp. from the Mulde Beds at Djupvik. □F. Dorsal view of steinkern (anterior to the right) showing muscular impressions (same lettering as in Fig. 4, v = visceral attachment muscles, cf. Fig. 10), RMMo 24279, ×3.2. □G. Lateral right view of steinkern, note pallial sinus (at arrows), RMMo 21931, ×4.1. □H. Left view of articulated specimen (holotype), note posterior sulcus (at arrows), RMMo 15618, ×4.3. □I. Right view of steinkern showing muscular impressions (same lettering as in Fig. 4, cf. Fig. 10; same specimen as F), ×2.1. □J. Dorsal view of holotype (anterior to the left), note attached bryozoan colony in posterior part of right valve (outline marked with ink), ×3.3. □K. Hinge of left valve (posterior teeth damaged), note ligament nymph (between arrows), RMMo 15617, ×9.7. □L–P. *Palaeostraba baltica* (Liljedahl, 1984) from the Halla Beds at Möllbos 1. □L. Ventro-lateral view of holotype (left valve), note internal septa (at white arrows) and umbonal depression (at black arrow), SGU Type 3498, ×4.0, sample G78-2LL. □M. Lateral view of holotype, ×4.3. □N. External view of articulated specimen (anterior to the right; posterior part of right valve deformed), SGU Types 3638/3639, ×4.3, sample G79-79LJ. □O. Lateral view of holotype, note posterior sulcus (at arrows), ×4.3. □P. Hinge of right valve, SGU Type 3607, ×16.0, sample G79-78LJ. A–K are preserved as calcium carbonate and the remaining specimens are silicified.

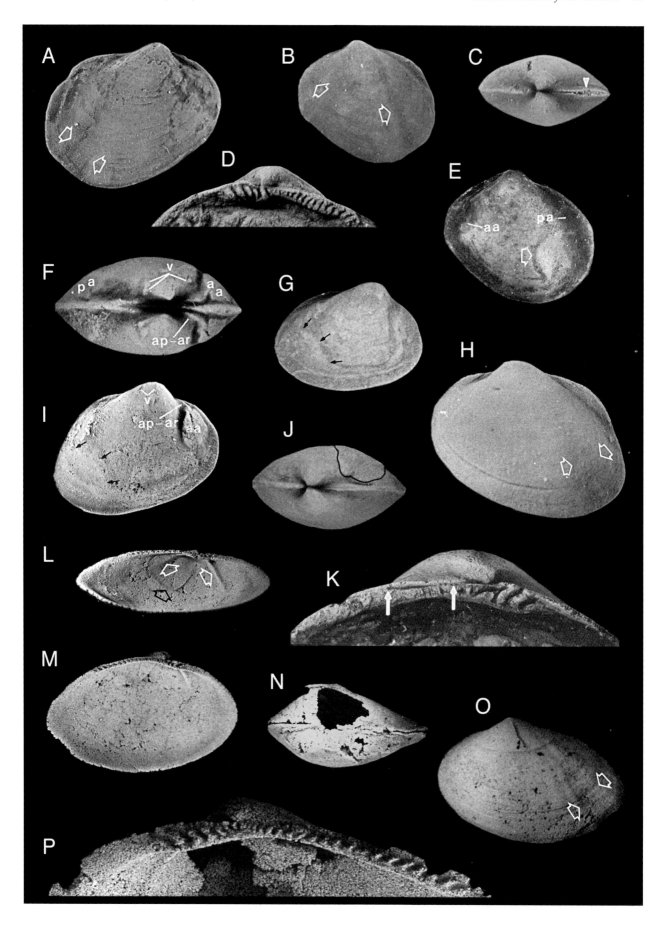

11.5 mm long containing 4–5 teeth; posterior hinge plate long, narrow with chevron-shaped teeth having apices towards beak, a specimen 11.5 mm long containing 20–30 teeth (Fig. 34K); anterior adductor muscle scar large, subcircular, deeply impressed (aa in Fig. 10; Fig. 34F, I); posterior adductor muscle scar subcircular, somewhat smaller and not so deeply incised as anterior one (pa in Fig. 10; Fig. 34F, I); pedal muscle scars between anterior adductor muscle scar and umbonal cavity and in umbonal cavity (v in Fig. 10; Fig. 34F, I); evident pallial sinus (ps in Fig. 10; Fig. 34I).

Remark. – *Ekstadia kellyi* has a more or less evident external diagonal sulcus in one end containing also a pallial sinus, indicating the presence of siphons. This end is considered to be posterior.

Soot-Ryen (1964, p. 512) denoted the holotype of *Ekstadia kellyi* as *Nuculoidea* sp. A, considering it to have opisthogyrate beaks, i.e. a reversed orientation to that presented herein. The material of Soot-Ryen (1964) originally determined and labelled *Ekstadia tricarinata* contained a few specimens of *Ekstadia kellyi*.

The following muscular impressions have been observed (Fig. 10): anterior adductor muscle scar (aa), posterior adductor muscle scar (pa), anterior pedal protractor or retractor muscle scar (ap+ar), visceral attachment muscle scars (v) and pallial line with shallow sinus (ps).

Comparisons. – See discussion of *Ekstadia tricarinata*.

Occurrence. – Wenlockian (Homerian) Mulde Beds at Djupvik and Ludlovian Hemse Beds at Petesvik.

Genus *Palaeostraba* Liljedahl, 1984

Systematic position. – *Palaeostraba* seems to be closely related to the Devonian *Straba* Prantl & Růžička, 1954, type genus of the subfamily Strabinae, placed in the family Ctendodontidae by Prantl & Růžička (1954, p. 10). However, since *Palaeostraba* is assumed to be siphonate, has an external opisthodetic ligament, lacks resilifer but probably also had an additional internal part of the ligament, it is placed in the family Malletiidae of the Nuculanacea.

Type and only species. – *Palaeostraba baltica* Liljedahl, 1984.

Diagnosis. – Malletiid with slightly elongated posterior end, beaks slightly prosogyrate (Fig. 34N, O), opisthodetic external ligament and possibly also an additional internal part of the ligament (Fig. 34P); distinct anterior internal septum and less conspicuous posterior septum (Fig. 34L); central tooth and socket extremely small, as are adjoining anterior and posterior teeth (Fig. 34P); oblique external sulcus from postero-umbonal slope to postero-ventral margin (Fig. 34O).

Palaeostraba baltica Liljedahl, 1984

Fig. 11:13, 12:5,15, 26, 34L–P

Synonymy. – □1984a *Palaeostraba baltica* n.sp. – Liljedahl, pp. 13–19, Figs. 4, 6–9. □1984b *Palaeostraba baltica* Liljedahl – Liljedahl, pp. 63–64; Figs. 6–7. □1984c *Palaeostraba baltica* Liljedahl – Liljedahl, pp. 2, 8–9. Fig. 1.

Holotype. – SGU Type 3498, a single left valve. Length = 11.8 mm, height = 7.5 mm, Fig. 34L–M, O.

Type stratum. – Halla Beds, Upper Wenlockian.

Type locality. – Möllbos 1, Gotland.

Material. – About 30 silicified specimens.

Diagnosis. – Same as for the genus.

Remarks. – For detailed description see Liljedahl (1984, pp. 13–19). The diagonal sulcus of *Palaeostraba baltica* is regarded as being in the posterior end.

The specimens examined are all exclusively silicified. The external sulcus of *Palaeostraba baltica* indicates the presence of siphons (cf. corresponding feature in modern siphonate bivalves). This assumption is supported by ecological evidence (see 'Ecology and faunal associations').

The two series of hinge teeth are continuous, and immediately posterior to the beak below the proximalmost tooth in the posterior hinge plate there is a non-denticulate area (Fig. 34P), possibly being the site of an internal part of the ligament.

The muscular impressions of this species are few and extremely faint; the anterior adductor muscle scar is larger than the posterior one. An evident depression in the umbonal cavity (Fig. 34L) is interpreted as the site of the visceral mass (cf. reconstruction of soft parts of *Caesariella lindensis* in Fig. 8B herein).

Dimensions of holotype. – Length = 11.6 mm, height = 7.5 mm, width = 2.3 mm, H/L = 0.65, 2W/L = 0.40.

Occurrence. – Late Wenlockian (Homerian) Halla Beds at Möllbos 1, Gotland.

Genus *Nuculites* Conrad, 1841

Type species. – *Nuculites oblongata* Conrad, 1841.

Nuculites solida (Soot-Ryen, 1964)

Figs. 11:14, 14:2, 15,26, 35A–C

Synonymy. – □1964 *Palaeoneilo solida* n.sp. – Soot-Ryen, p. 501, Figs. 2:3,5

Holotype. – RMMo 150355, a left valve, Fig. 35A, C.

Type stratum. – Probably Burgsvik Beds, Whitcliffian, Late Ludlovian.

Type locality. – Probably Burgsvik, Gotland.

Material. – Three specimens with shells preserved as recrystallized calcite.

Diagnosis. – Shell large with expanded posterior end, external opisthodetic ligament, diagonal sulci, beaks prosogyrate, hinge teeth continuous below beak, anterior teeth larger and less numerous than posterior ones, strong internal septum in anterior part.

New description. – Shell large, arciform, subovate, gibbous, inequilateral, external opisthodetic ligament (Fig. 35C), diagonal shallow sulci from umbo to posterior part of ventral margin (Fig. 35A–B); umbo low, beaks prosogyrate, close together, slightly raised above dorsal margin, in anterior half of shell (Fig. 35C); no lunule or escutcheon; shell surface smooth with concentric growth lines; dorsal margin slightly convex; anterior margin rounded, slightly truncated; posterior margin obliquely truncated; ventral margin straight, sinuated in posterior part where sulcus emerges; no ligament observed; hinge teeth in an uninterrupted series, anterior hinge plate short, broad with strong chevron-shaped teeth, apices towards beak, in holotype (approximately 25 mm long) containing 11 teeth; posterior hinge plate long, narrow in distal part, with chevron-shaped teeth having their apices towards beak, in remaining part lamellar, holotype containing approximately 30 teeth; teeth extremely small below beak, posterior teeth smaller than anterior teeth (Fig. 35C); hinge angle approximately 150°; no resilifer; deep, thick internal septum (Fig. 35A).

Dimensions of holotype. – Length = 25.4 mm, height = 15.0 mm, width = 7.0 mm, H/L = 0.59, 2W/L = 0.55.

Remarks. – The diagonal sulcus of *Nuculites solida* is regarded as being in the posterior end.

Systematic position. – Soot-Ryen (1964, p. 501) placed this species in the genus *Palaeoneilo* Hall, 1869, most probably because of its diagonal sulci. However, this species also exhibits a strong internal septum, a feature present only in *Nuculites* Conrad, 1841 (low, thin internal septa, or ridges are present in a few additional nuculoids, e.g., *Palaeostraba*). *Palaeoneilo* differs from *Nuculites* in form and in the absence of an internal septum (cf. discussion in Hall 1885, pp. XXVI, XXVII; cf. also description of the type species of *Nuculites* and *Palaeoneilo*, respectively, in McAlester 1968, pp. 37, 41).

Owing to the presence of a robust internal septum in the present species, it seems to be more closely related to *Nuculites* Conrad than to *Palaeoneilo* Hall.

Occurrence. – Ludlovian (Whitcliffian) Burgsvik Beds, probably at Burgsvik.

Nuculites sp. A sensu Soot-Ryen, 1964

Figs. 15, 26, 35D

Synonymy. – □1964 *Nuculites* sp. A – Soot-Ryen, p. 500, Fig. 2:6

Material. – Only one internal mould in limstone and some additional fragments on the same slab.

Diagnosis. – Shell medium-sized with expanded posterior end, umbo low, conspicuous internal septum.

For a detailed description see Soot-Ryen (1964, p. 500).

Dimensions. – Length of specimen ca. 16 mm, height 9 mm (H/L = 1,8).

Occurrence. – Ludlovian Hemse Beds, probably from the Etelhem Limestone, locality unknown.

Nuculites sp. B sensu Soot-Ryen, 1964

Figs. 14:5, 15, 26, 35E, G

Synonymy. – □?1909 *Cuculella ovata* Sowerby – Moberg & Grönwall, p. 37, Pl. 2:20–21. □1964 *Nuculites* sp. B – Soot-Ryen, p. 500, Pl. 3:1–2.

Material. – One specimen of a left valve with shell preserved as recrystallized calcite, RMMo 15823 (Fig. 35E). Length = 22.0 mm, height = 13.0 mm, width = 2.4 mm, H/L = 0.59, 2W/L = 0.44. Two internal moulds of right valves of fine-grained limestone, RMMo 21977 (Fig. 35G) and RMMo 15620, one incomplete internal mould of a right valve of fine-grained limestone, RMMo 21982, and one internal mould with dorsal part with umbo preserved as recrystallized calcium carbonate, RMMo 21518.

Diagnosis. – Shell large with expanded posterior end, umbo small, faint diagonal sulcus, shell surface with coarse, regular concentric lamellae, internal septum.

For a detailed description see Soot-Ryen (1964, p. 50).

Remark. – The internal septum of the specimen with preserved shell is discernible owing to shell collapse (Fig. 35E). In the remaining specimens, consisting of internal moulds, the septum is clearly observable as a split in the mould (Fig. 35G). Possibly this species corresponds to *Cuculella ovata* Sowerby, described by Moberg & Grönwall (1909) from the Silurian of Scania. The Scanian species seems to be a little more gibbous than *Nuculites* sp. B, but differences in valve convexity may be due to state of preservation.

Occurrence. – Ludlovian (Eltonian) Hemse Beds at Östergarn and Lau kanal and Whitcliffian Burgsvik Beds at Gansviken and at the shore of Grötlingbo.

Family Nuculanidae Adams & Adams, 1858

Diagnosis (McAlester 1969). – Ligament partially internal, resilifer present.

Systematic position. – Palaeozoic nuculoids with attenuated posterior part are in older literature usually referred to the extant genus *Nuculana* Link (= *Leda* Schumacher). McAlester (1962) argued that fossil taxa ought not to be assigned to extant ones unless clear evidence of conchological similarity could be accounted for. Because of the obliterated original mineralogy of Palaeozoic bivalve shells, he suggested that *Nuculana*-like Palaeozoic bivalves belong to an undescribed taxon or taxa.

Posteriorly extended Palaeozoic nuculoids may exhibit different internal characteristics. For example *Nuculites* Conrad, 1841, has an anterior internal septum, *Ditichia* Sandberger, 1891, two internal septa, and *Phestia* Chernyshev, 1951, internal umbonal ridge or ridges.

Since only the hinge of the two present nuculanids is known, they are not possible to place into any genus or genera.

'*Nuculana*' *oolitica* (Soot-Ryen, 1964)

Figs. 11:15, 14:3, 15, 26, 35F, J–K

Synonymy. – □?1858 *Leda pandoriformis* Stevens – Stevens, p. 262. □1883 *Leda* (*Nuculana*) *ohioensis* Hall – Hall, Pl. 47:49–50. □?1885 *Leda pandoriformis* (Stevens) – Hall, Pl. 47:49–50. □*non* 1897 *Ctenodonta securiformis* (Goldfuss) – Grönwall, p. 220 (fossil list). □*non* 1909 *Ctenodonta securiformis* (Goldfuss) – Grönwall *in* Moberg & Grönwall, p. 38, Pl. 2:18,19. □1964 '*Nuculana*' *oolitica* n.sp. – Soot-Ryen, p. 513.

Material. – Only one specimen known so far, a right valve with shell preserved as recrystallized calcite.

For a detailed description see Soot-Ryen (1964, p. 513).

Remark. – The rostrate end of '*Nuculana*' *oolitica* is considered to be the posterior.

The number of hinge teeth present in the specimen available indicates that it is an adult (cf. discussion in Liljedahl 1983, pp. 22–36).

Dimensions of specimen. – Length = 8.0 mm, height = 4.7 mm, width = 1.7 mm, H/L = 0.59, W/L = 0.21.

Comparisons. – '*Nuculana*' *oolitica* closely resembles *Leda pandoriformis* Stevens, 1858, from the Silurian of North America and is most probably conspecific with this species. '*Nuculana*' *oolitica* is also similar to *Nuculana securiformis* Goldfuss, 1834, from the Silurian of Scania, Sweden. '*Nuculana*' *oolitica* has a slightly higher number of teeth in the posterior hinge plate than in the anterior one while *Nuculana securiformis* has almost the double amount of teeth in the

anterior hinge plate as compared with the posterior one (see Beushausen 1895, p. 60).

'*Nuculana*' sp. A Soot-Ryen, 1964

Figs. 11:16, 13:3, 15, 26, 35H–I

Synonymy. – □1964 '*Nuculana*' sp. A – Soot-Ryen, p. 514.

Remarks. – This taxon is represented by one complete articulated specimen only (RMMo 15381) with the shell preserved as recrystallized calcite. The existence of taxodont teeth may be observed by transmitted light. The rostrate end is considered to be the posterior.

For a detailed description se Soot-Ryen (1964, p. 514).

Dimensions of only specimen. – Length = 6.9 mm, height = 3.9 mm, 2×width = 1.6 mm, H/L = 0.56, 2W/L = 0.23.

Comparisons – The present taxon superficially resembles several Palaeozoic nuculoids such as *Ctenodonta longa* Ulrich, 1897, p. 590, Pl. 37:30–31, *Leda perdentata* Barrande, 1881, Pl. 270:2 and *Leda decurtata* Barrande, 1881, Pl. 70:4.

Occurrence. – Wenlockian (Homerian) Mulde Beds at Djupvik, Gotland.

Fig. 35. □A–C. *Nuculites solida* (Soot-Ryen, 1964). □A. Lateral view of holotype (left valve) showing internal septum where shell material has been worn off (at arrow), RMMo 150355, ×1.9, Burgsvik Beds, probably at Burgsvik. □B. Lateral view of left valve, RMMo 150350, ×2.4, Burgsvik Beds at Burgsvik. □C. Hinge of holotype, ×3.2. □D. *Nuculites* sp. A (Soot-Ryen, 1964). Right valve of internal mould showing internal septum (preserved as a slit), RMMo 21983, ×2.8, Hemse Beds, locality unknown. □E. *Nuculites* sp. B (Soot-Ryen, 1964). External view of incomplete left valve, RMMo 15823, ×2.9, Burgsvik Beds at Gansviken. □F. '*Nuculana*' *oolitica* (Soot-Ryen, 1964). Dorsal view of holotype (right valve), RMMo 150277, ×4.9, Burgsvik Beds at Gansviken. □G. *Nuculites* sp. B (Soot-Ryen, 1964). Right view of internal mould, showing slit representing internal septum, RMMo 21977, ×2.0, Burgsvik Beds, probably at Gansviken. □H–I. '*Nuculana*' sp. A (Soot-Ryen, 1964). □H. Dorsal view of articulated specimen (holotype), anterior to the right, RMMo 15381, ×7.4, Mulde Beds at Djupvik. □I. Right view of holotype, ×7.5. □J–K. '*Nuculana*' *oolitica* (Soot-Ryen, 1964). J. Hinge of holotype, ×11.2. □K. Lateral view of holotype, ×5.7. □L. *Tancrediopsis gotlandica* (Soot-Ryen, 1964). Dorsal view of articulated specimen, note opistodetic ligament, anterior of shell to the right, RMMo 150326, ×3.1, Mulde Beds at Djupvik. □M. *Tancrediopsis solituda* (Soot-Ryen, 1964). Lateral left view of articulated specimen (holotype), RMMo 15321, ×3.3, Mulde Beds at Djupvik. □N–O. *Tancrediopsis gotlandica* (Soot-Ryen, 1964). □N. Lateral right view of articulated specimen (lectotype), RMMo 149880, ×1.7, Mulde Beds at Djupvik. □O. Lateral right view of articulated specimen, LO6265t, ×1.1, Mulde Beds at Mulde Brickyard. □P. *Tancrediopsis gotlandica* or *Tancrediopsis solituda*. Hinge of right valve, RMMo 150278m ×4.3, Visby Beds?, Fårö, locality unknown.

?Superfamily Nuculanacea Adams & Adams, 1858

?Family Malletiidae Adams & Adams, 1858

Genus *Tancrediopsis* Beushausen, 1895

Synonymy. – *Praectenodonta* Philip, 1962, p. 226; *Gotodonta* Soot-Ryen, 1964, p. 502.

Type species. – *Gotodonta gotlandica* (Soot-Ryen, 1964), original designation *Nucula sulcata* Hisinger, 1841, Pl. 40:2.

Discussion. – When Beushausen (1895, pp. 70, 94) proposed the new genus *Tancrediopsis* for the reception of some species previously included in *Ctenodonta s.s.* Salter, 1852, he suggested two type species, viz. *Ctenodonta contracta* Salter, 1859, (p. 37) and *Nucula sulcata* Hisinger, 1841 (Pl. 402a–b).

The first subsequent designation of one of these species as the type species of *Tancrediopsis* was made by Cossman (1897, p. 94) who chose *Ctenodonta contracta* Salter, 1859 [(McAlester, 1963, considered the correct name of *Ctenodonta contracta* Salter to be *Tancrediopsis cuneata* (Hall, 1856)]. However, the original diagnosis of *Tancrediopsis* does not agree with the actual morphology of the type species chosen by Cossman but rather with that of *Nucula sulcata* Hisinger, 1841 (*non* Bronn, 1832). Consequently, *Nucula sulcata* Hisinger, 1841, is suggested to be the type species of *Tancrediopsis*.

Soot-Ryen (1964, p. 504) pointed out that *Nucula sulcata* Hisinger, 1841, is a junior homonym of *Nucula sulcata* Bronn, 1832. And, since there were no available synonyms for this species, she renamed it *Gotodonta gotlandica*. Accordingly, the correct name of the type species of *Tancrediopsis* is *Tancrediopsis gotlandica* (Soot-Ryen, 1964).

When Soot-Ryen (1964, p. 502) proposed the new genus *Gotodonta*, she was apparently not aware of the existence of *Praectenodonta* Philip, since she included the very same species in *Gotodonta* as Philip did in his *Praectenodonta*. This suggestion is supported by the fact that some years after 1964, Soot-Ryen changed the labels of the specimens of *Gotodonta gotlandica* to *Praectenodonta gotlandica*, in the collections of the Museum of Natural History, Stockholm. *Gotodonta* Soot-Ryen, 1964, is regarded as a junior synonym of *Tancrediopsis* Beushausen, 1895.

The second subsequent designation of a type species of *Tancrediopsis* was made by McLearn (1924). He chose as type species *Tancrediopsis subcontracta* Beushausen, 1895, since this was actually the species upon which Beushausen based the diagnosis of the genus. *Tancrediopsis subcontracta* is, however, not valid as the type species of *Tancrediopsis*, because it is preceded by Cossman's designation of *Ctenodonta contracta*, in 1897.

Philip (1962) considered *Ctenodonta contracta* and *Nucula sulcata* as fairly typical for *Ctenodonta*, actually closely resem-

bling the type species *Ctenodonta nasuta* Hall, 1847. He therefore regarded *Tancrediopsis* Beushausen as a subjective synonym of *Ctenodonta* Salter, 1852.

It may be true that *Ctenodonta contracta* does not differ so much from *C. nasuta* that its affiliation to another genus is justified. However, the second type species proposed by Beushausen, *Nucula sulcata* Hisinger, differs considerably from the typical *Ctenodonta* and thus warrants incorporation into another genus. *Nucula sulcata* Hisinger deviates from *Ctenodonta s.s.* in being conspicuously rostrate posteriorly, in having a distinct umbonal ridge forming a depressed escutcheon which reaches the postero-ventral margin, and finally in having a shell surface with a conspicuous growth pattern of concentric ribs.

As a consequence of Cossman's unfortunate choice of *Ctenodonta contracta* as the type species of *Tancrediopsis* instead of *Nucula sulcata*, Philip (1962, p. 226) erected the new genus *Praectenodonta*. He based his diagnosis of *Praectenodonta* upon the two species *Palaeoneilo raricostae* Chapman, 1908, and *Palaeoneilo victoriae* Chapman, 1908. *Palaeoneilo raricostae* was established as type species by original designation (Philip, 1962, p. 226).

The diagnosis of *Praectenodonta* Philip, 1962, p. 226, reads: 'Rostrate forms . . . with surface ornamentation of broadly spaced, strong growth lamellae', both features being diagnostic for *Tancrediopsis* Beushausen. The spacing of surface ribs may be regarded as specific and not generic characteristic. However, since the main characteristics for *Praectenodonta* are diagnostic for *Tancrediopsis*, *Praectenodonta* Philip is regarded as a junior synonym of *Tancrediopsis* Beushausen.

Tancrediopsis gotlandica (Soot-Ryen, 1964)

Figs. 11:17, 13:7, 15, 26, 35L, N–O, P(?)

Synonymy. – ☐*non* 1832 *Nucula sulcata* Bronn – Bronn, p. 617. ☐1841 *Nucula sulcata* (Nob.) – Hisinger, p. 66, Pl.40:2a, b. ☐*non* 1867 *Ctenodonta sulcata* Hisinger – Murchison, p. 530. ☐1880 *Ctenodonta sulcata* Hisinger – Lindström *in* Angelin & Lindström, p. 18. ☐1895 *Tancrediopsis subcontracta* Beushausen – Beushausen, p. 94, Pl. 8:14–16. ☐1895 *Ctenodonta sulcata* Hisinger – Beushausen, p. 70. ☐1921 *Ctenodonta sulcata* (Hisinger) – Hede, pp. 48, 95. ☐1964 *Gotodonta gotlandica* nom.nov. – Soot-Ryen, p. 504, Fig. 3:5–7.

Lectotype. – RMMo 14988, Fig. 35N.

Type stratum. – Mulde Beds, Late Wenlockian.

Type locality. – Djupvik 1, Gotland.

Material. – More than 200 specimens.

Emended diagnosis. – Shell medium-sized, posterior extremity almost rostrate (Fig. 35N–O); beaks small, opisthogyrate

(Fig. 35L), sharp posterior umbonal ridge, distinct lunule with external opisthodetic ligament (Fig. 35L); shell surface with concentric ribs; subequal hinge plates (Fig. 35P), hinge teeth continuous below beak.

For a detailed description see Soot-Ryen (1964, pp. 504–505).

Dimensions. – Lectotype length 15.5 mm, height 8.2 mm, width 5.6 (both valves). H/L = 0.53, W/L = 0.36 (both valves).

Remark. – *Tancrediopsis gotlandica* has a conspicuous external ligament in the short part of the valve, which furthermore is rostrate, suggesting this part to be posterior.

Discussions of closely related forms are accounted for by Soot-Ryen (1964, pp. 503–505).

Occurrence. – Wenlockian Upper Visby Beds at Vialms, Fleringe, Liksarve Tofta; Slite Beds on Fårö; Mulde Beds at Djupvik 1, Gotland.

Tancrediopsis solituda (Soot-Ryen, 1964)
Figs. 11:18, 13:1, 15, 26, 35M, P(?)

Synonymy. – ☐1964 *Gotodonta solituda* n.sp. – Soot-Ryen, p. 507, Fig. 3:8.

Holotype. – RMMo 15391, a complete, articulated specimen preserved as recrystallized calcite, Fig. 35M.

Material. – Only the holotype.

Type stratum. – Mulde Beds, Late Wenlockian.

Type locality. – Djupvik 1, Gotland.

Diagnosis. – Shell small, posterior extremity almost rostrate; beaks small, slightly opisthogyrate, sharp posterior umbonal ridge; ventral margin slightly sinuated in posterior end; no lunule; external opisthodetic ligament; shell surface with coarse concentric lamelliform ridges.

For a detailed description, see Soot-Ryen (1964, p. 507).

Dimensions of holotype. – Length = 13.6 mm, height = 7.6 mm, Width = approximately 3.0 mm (one valve), H/L = 0.56, W/L = 0.22 (both valves).

Remark. – *Tancrediopsis solituda* has a conspicuous external ligament in the short part of the valve, which furthermore is rostrate, suggesting this part to be posterior.

Tancrediopsis solituda differs from *T. gotlandica* in having a much coarser and wider spaced sculpture.

Occurrence. – Wenlockian Slite Beds at Biskops, and Bolarve hällar, and Mulde Beds at Djupvik 1, Gotland.

Subclass Isofilibranchia Iredale, 1939

Emended diagnosis. – Equivalve *to subequivalve*, inequilateral, *generally* byssate, opisthodetic elongate ligament, hinge eden-

tulous or with one or several cardinal teeth, no lateral teeth, anisomyarian musculature (emendation in italics).

Order Mytiloida Férussac, 1822

Diagnosis. – Same diagnosis as the subclass.

Superfamily Mytilacea Rafinesque, 1815

Diagnosis. – Same as Isofilibranchia and Mytiloida

Family Modiomorphidae Miller, 1877

Diagnosis. – Modioliform and goniophoriform mytilaceans with expanded posterior end, no resilial ridge to support ligament.

Subfamily Modiomorphinae n.subf.

Diagnosis. – Modiomorphidae with hinge teeth.

Type genus. – *Modiomorpha* Hall, 1869

Genera. – *Aleodonta* n.gen.; *Callodonta* Isberg, 1934; *Colpomya* Ulrich, 1893; *Eurymyella* Williams, 1912; *Goniophora* Phillips, 1848; *Mimerodonta* n.gen.; *Modiolodon* Ulrich, 1893; *Modiomorpha* Hall, 1869; and *Radiatodonta* Dahmer, 1921.

Genus *Modiodonta* Liljedahl, 1989

Systematic position. – The most closely related genus seems to be *Colpomya* Ulrich, 1893, which differs from *Modiodonta* in having a characteristic subrhomboidal shape with sharp anterior and posterior umbonal ridges separated by a conspicuous mesial depression and a distinctly sinuated ventral margin. *Mimerodonta* and *Aleodonta* also seem to be closely related. From these, *Modiodonta* differs as follows. *Modiodonta* has closely fitting dorsal margins, a convex hinge line, a true hinge plate with one cardinal tooth in the right valve and a socket in the left. *Aleodonta* has overlapping dorsal margins and lacks hinge plate in the ordinary sense, while *Mimerodonta* has a straight hinge line, lacks hinge plate and has one cardinal tooth and two shallow sockets in the right valve and one deep socket flanked by two tooth-like reinforcements in the left.

Other modiomorphid genera differ more from *Modiodonta*. Thus, *Modiomorpha* Hall & Whitfield, 1869, possesses a conspicuous external diagonal sulcus. The resulting sinuous

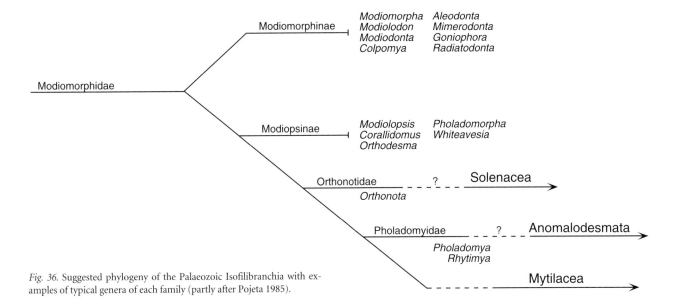

		Modiomorpha	Aleodonta
	Modiomorphinae	Modiolodon	Mimerodonta
		Modiodonta	Goniophora
		Colpomya	Radiatodonta
Modiomorphidae			
		Modiolopsis	Pholadomorpha
	Modiopsinae	Corallidomus	Whiteavesia
		Orthodesma	

Fig. 36. Suggested phylogeny of the Palaeozoic Isofilibranchia with examples of typical genera of each family (partly after Pojeta 1985).

ventral margin contrasts markedly against the convex one of *Modiodonta*. The hinge of *Modiomorpha* consists of a robust tooth in the left valve, while the hinge dentition of *Modiodonta* is a tooth in the right valve.

Modiodonta differs from *Modiolodon* Ulrich, 1893, by being much smaller and more elongated and in having a more pronounced umbo and umbonal ridge. *Modiodonta* has one tooth in the right valve only, while *Modiolodon* exhibits 1–3 teeth in each valve.

Since *Modiodonta* is characterized by a distinct hinge tooth, it is easily distinguished from all edentulous forms, such as *Modiolopis* Hall, 1847, and *Whiteavesia* Ulrich, 1893.

Type and only species. – *Modiodonta gothlandica* (Hisinger, 1831).

Emended diagnosis. – Equivalve, inequilateral, diagonally subelliptical, margins fitting; more or less prominent umbonal ridge, beaks subanterior, shell surface smooth with irregular concentric growth lines; no escutcheon, lanceolate lunule; internal linear ligament; dorsal margin entire below beak; convex hinge line; moderately strong hinge plate with one cardinal tooth in right valve, corresponding socket in left; anterior adductor muscle scar deep, smaller than posterior one, a number of accessory muscle scars in anterior region, one of which is the impression of a strong byssal–pedal retractor; generally with conspicuous posterior byssal–pedal retractor in hypertrophied posterior adductor muscle scar; integripalliate.

Modiodonta gothlandica (Hisinger, 1831)

Figs. 19A, F, 23:1, 27, 37A–G, 38

Synonymy. – □*1827 *Mytilus –* Hisinger, p. 323, Pl. 7:7. □v*1828 *Mytilus* Linn. – Hisinger, p. 220, Pl. 7:7. □1831a

Modiola gothlandica H. – Hisinger, p. 117. □1831b *Modiola gothlandica* H. – Hisinger, p. 15. □v*1837 *Modiola? Nilssoni –* Hisinger, pp. 61, 123, Pl. 18:13. □?1839 *Modiola antiqua –* Sowerby, p. 628, Pl. 13:1. □1840 *Modiola Nilssoni His.* – Hisinger, p. 63. □1841 *Modiola? Nilssoni His.* – Hisinger, p. 56. □?1855 *Modiolopsis Nilsoni* (His. sp.) – McCoy, p. 267, Pl. 1I:21. □non 1865 *Modiolopsis Hall* Nilssoni His. – Kjerulf, p. 11, Fig. 27. □non 1880 *Modiolopsis Nilssoni* His. – Lindström, *in* Angelin & Lindström, p. 59, Pl. 19:5,6. □non 1880 *Ptychodesma Nilssoni* Hisinger – Lindström, *in* Angelin & Lindström, p. 18. Pl. 2:21,22. □?1882 *Ptychodesma Gothlandicum* His. – Lindström, p. 18. □?1885 *Ptychodesma Gothlandicum* His. – Lindström, p. 12. □?1888 *Ptychodesma gothlandicum* His. – Lindström, p. 12. □?*non* 1921 *Modiolopsis Nilssoni* (His.) – Hede, pp. 51, 95. □?*non* 1921 *Modiolopsis Nilssoni* (His.) – Hede, pp. 20, 21, 54. □*non* 1929 *Modiolopsis Nilssoni* Hisinger – Quenstedt, *pp.* 40–46, Pl. 1:22–25. □1989 *Modiodonta gothlandica* (Hisinger, 1831) – Liljedahl, pp. 313–318, Figs. 1–5.

Holotype. – Internal mould of an articulated specimen with slightly opened valves, RMMo 149878 (Fig. 37D, G). Originally figured by Hisinger 1827, Pl. 7:7.

Type stratum. – Mulde Beds, Wenlockian.

Type locality. – Eksta Parish, Djupvik (for the extent of this old composite locality designation, see Jaanusson 1986, p. 13, 14).

Material. – Fifteen internal moulds of articulated specimens, two articulated specimens, and one single valve with preserved shell.

Diagnosis. – Shell medium-sized (maximum length ca. 30 mm), subelliptically elongate (Fig. 37B, F–G) with prominent oblique umbonal ridge having sulcus at proximal part of umbo (Fig. 37B), beaks prosogyrate, close together; length almost twice the height; strong hinge plate with a single robust cardinal tooth in right valve, corresponding socket in left (Fig. 37A, C, E); deep anterior adductor muscle scar (Fig. 37A, C) smaller than posterior scar; one deep anterior byssal–pedal retractor muscle scar and several additional scars in umbonal region (Fig. 37A, C, E).

Description. – For description, see Liljedahl (1989). L = 19.7–30.6 mm; L/H = 1.85; H/2W = 1.99.

Comparisons. – *Modiodonta gothlandica* differs from ?*Colpomya audae* and ?*Colpomya lokei* in having raised dorsal margins, a less pronounced anterior lobe, and an evenly rounded posterior end.

Modiodonta gothlandica differs from *Colpomya hugini* by its less pronounced anterior lobe, more rounded lateral outline, and larger size.

Modiodonta gothlandica differs from ?*Colpomya balderi* by its more elongate shape and slightly more accentuated umbonal ridge.

Superficially *Modiodonta gothlandica* also resembles a number of modiomorphids with unknown interior. For example, *Modiolopsis? consimilis* Ulrich 1897 (Pl. 42:17) from the Ordovician of Minnesota resembles *Modiodonta gothlandica* in lateral outline but is shorter, more gibbous, and has its maximum convexity at mid-length of the shell.

The Wenlockian *Modiola antiqua* Sowerby (1839, p. 628, Pl. 13:1) is shorter than *Modiodonta gothlandica*, has an undefined umbonal ridge, straight posterior dorsal margin and a sinuous ventral margin.

McCoy (1855, p. 267, Pl.1I:21) described and illustrated *Modiolopsis nilssoni* (Hisinger) from the Ludlovian of Britain. It differs externally from *Modiodonta gothlandica* in being decidedly shorter and in having a faint sinus in the middle of the ventral margin, an undefined umbonal ridge, a long, almost straight posterior dorsal margin and a less developed umbo than *Modiodonta gothlandica*.

Kjerulf (1865, p. 11, Fig. 27) reported *Modiolopsis nilssoni* (Hisinger) from the Middle Ordovician of Norway and compared it to one specimen illustrated by Hisinger (1837, Pl. 18:13). That specimen, however, is not *Modiodonta gothlandica* (see Liljedahl 1989). He also referred to *Modiolopsis modiolaris* (Conrad) with which his specimen seems to be more closely related, judging from the external features.

Quenstedt (1929) described *Modiolopsis Nilssoni* Hisinger from Upper Silurian deposits of Spitsbergen. He stated that the hinge apparently was edentulous and the muscular impressions weak. In three of the Spitsbergen specimens there is a distinct sinus in the ventral margin.

Ecology and habitat. – *Modiodonta gothlandica* exhibits characteristics typical of endo-byssate bivalves, e.g., elongate compressed shell, reduced anterior end, maximum convexity above mid-height, and both posterior and anterior byssal retractor muscle scars. The angle between the line through the ligament and that passing through the centre of both adduc-

tor muscle scars is almost the same as in the Recent *Modiola demissus* (cf. Stanley 1972, p. 170, Fig. 3). The hypertrophy of the posterior adductor muscle scar, containing the impression of a posterior byssal retractor (Fig. 19F, 37F)) is not as pronounced as in the extant epibyssate *Mytilus edulis* or even in the endobyssate *Modiolus modiolus*, *Modiolus demissus*, and *Brachidontes citrinus* (cf. Stanley 1972, Fig. 4). Thus, the posterior byssal retractor muscle scar of *Modiodonta gothlandica* is situated more posteriorly than that of, e.g., *Modiolus demissus* and seems to have been furnished with a byssus of less strength. Extant endobyssate mytilids have a reduced foot and are not true burrowers (Stanley 1972, p. 171).

However, *Modiodonta gothlandica* also shows similarities to active burrowing species. Its anterior adductor muscle scar is comparatively larger than that of extant endobyssate mytilids. *Modiodonta gothlandica* exhibits an elaborate pattern of pedal muscle scars, including such of anterior pedal protractors and retractors as well as pedal elevators (Fig. 38).

Accordingly *Modiodonta gothlandica* exhibits morphological characteristics typical for both endobyssate and active burrowing forms. It is therefore suggested to have been an endo-byssate filter feeder while retaining an ability to be an effective burrower.

Another feature supporting the assumption of an active mode of life for this species is the distribution of the 'quick'- and 'catch' portions of the adductor muscle scars (See 'Morphology of the modiomorphid shell', section on musculature). In this species the 'quick' portion dominates (Fig. 19F), indicating the ability of rapid closure of the valves.

All specimens of *Modiodonta gothlandica* except one are articulated and were isolated from a strongly argillaceous calcilutite. The original sediment was a carbonate mud mixed with terrigenous clay, probably representing the natural habitat of this species.

Filter-feeding bivalves living in such a fine-grained sediment require a stable life position with the posterior part of the shell (where the inhalent and exhalent openings are situated) well above the sediment surface, to prevent clogging of the gills. An elongate shell like that of *Modiodonta gothlandica* could have achieved that by byssal fixation (Fig. 38B; cf. the extant endobyssate *Modiolus americanus* anchored to larger particles in fine-grained sediments – Savazzi 1984, p. 309, Text-fig. 1C).

Occurrence. – Wenlockian (Sheinwoodian) Slite Beds at Sanda; (Homerian) Mulde Beds at Djupvik, Gotland.

Genus *Colpomya* Ulrich, 1893

Type species. – *Colpomya constricta* Ulrich 1893, p. 659, Pl. 52:17–19, from the middle Ordovician (Shermanian) of North America.

Species. – *Colpomya constricta* Ulrich, 1893; *C. demissa* Ulrich, 1893; *C. hugini* n.sp.; *C. munini* n.sp.; ?*C. heimeri* n.sp.; ?*C.*

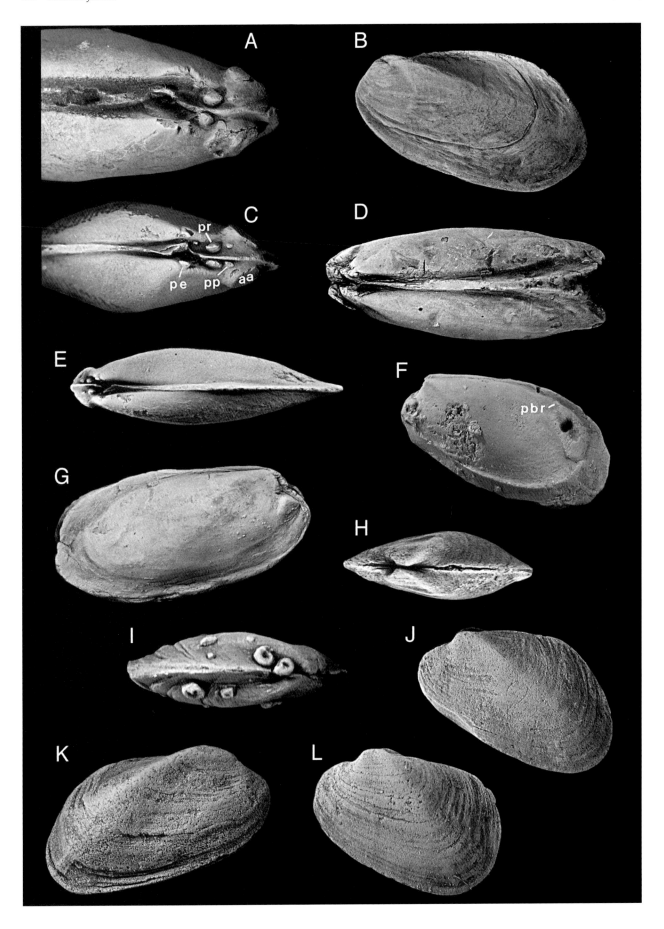

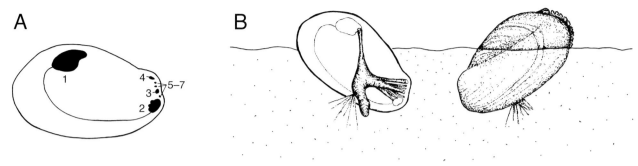

Fig. 38. Modiodonta gothlandica (Hisinger, 1831). A. Maximum number of muscular imprints in order of size. 1. Posterior adductor muscle scar. Note hypertrophied dorsoanterior part depicting a byssal retractor muscle scar, 2. Anterior adductor muscle scar with pedal accessory muscle scars, 3. Anterior byssal–pedal retractor muscle scar, 4. Pedal elevator muscle scar, 5–7. Pedal accessory muscle scars. B. Reconstruction of the foot and its muscles and suggested life position.

vaki n.sp.; ?*C. balderi* n.sp.; ?*C. ranae* n.sp.; ?*C. lokei* n.sp.; ?*C. friggi* n.sp.; ?*C. audae* n.sp.; possibly *C. concors* Reed, 1927.

Emended original diagnosis (Ulrich, 1894, p. 522). – Shell subelongate, oblique, equivalve *to subequivalve,* inequilateral, subrhomboidal or oval in outline, widest posteriorly, mesial sulcus normally pronounced, causing a flattening of the umbones and a sinus in ventral margin, umbonal ridge *more or less* prominent: hinge plate straight, thin posterior to beak, strong anterior to beak, hinge of right valve consisting of a cardinal tooth fitting into depression of left, posterior and anterior of this depression in the left valve is a *generally conspicuous,* strong process which is partly received in a shallow socket at the anterior side of the tooth in the right valve: *anterior adductor muscle scar deep, smaller than posterior one, a number of accessory muscle scars in anterior region, one of which being the impression of a strong byssal–pedal retractor, generally with conspicuous posterior byssal–pedal retractor in*

hypertrophied posterior adductor muscle scar; integripalliate. (Italics denote emendations.)

Colpomya hugini n.sp.
Figs. 16A, 17A, 19B, 23:2, 27, 37H–L, 39–40

Synonymy. – □1921 *Modiolopsis nilssoni* (Hisinger) – Hede, pp. 51, 55. □1927b *Modiolopsis nilssoni* (Hisinger) – Hede, pp. 20, 21, 54.

Derivation of name. – After one of the old Scandinavian god Woden's two ravens, *Hugin,* meaning Thought.

Holotype. – A right valve, RMMo 25411.

Type stratum. – Mulde Beds, Homerian, Wenlockian.

Material. – About 300 specimens, most of which are articulated.

Diagnosis. – Shell small, slightly inequivalve, normally right valve more convex than left, moderately inflated, conspicuous umbonal ridge more pronounced in right valve, sulcus in proximal part of umbonal ridge, no lunule; hinge plate with a single robust cardinal tooth in right valve and corresponding socket in left; strong anterior pedal–byssal retractor muscle scar.

External features. – Shell small (maximum length ca. 16 mm), obliquely subovate (Figs. 37J–L, 17J), inequilateral, moderately compressed (Fig. 37H–I), margins even, closely fitting, total length of shell about $1\frac{1}{2}$ of total height; shell surface smooth with irregular, faint concentric growth lines; beaks subanterior, small, close together, prosogyrate, not raised above commissure plane, slightly above dorsal margin; umbonal ridge more defined in right valve; maximum convexity at about mid-length and slightly above mid-height of shell; no escutcheon; no lunule; dorsal margin slightly convex to almost straight; anterior margin short, narrow and abruptly convex to almost pointed; posterior margin rounded, posterior end sometimes truncated: ventral margin long, slightly

Fig. 37. □A–G *Modiodonta gothlandica* (Hisinger, 1831). □A. Dorsal view of internal cast of articulated specimen showing muscular and hinge impressions in anterior part, RMMo 24942, ×5.3. □B. Lateral view of left valve of articulated specimen, RMMo 24939, ×2.5. □C. Dorsal view of internal cast of articulated specimen showing muscular and hinge impressions in the anterior part, RMMo 24944, ×4.2. □D. Dorsal view of internal cast of slightly opened articulated specimen, anterior facing right (holotype), RMMo 149878, ×2.6. (figured in Hisinger 1827, Pl. 7:7; 1828, Pl. 18:3). □E. Dorsal view of internal cast of articulated specimen, anterior facing left, same specimen as in C, ×3.1. □F. Lateral view of internal cast of articulated specimen, note pallial line and also hypertrophied posterior adductor muscle scar (arrow), anterior facing left, RMMo 24942, ×1.9. □G. Lateral view of holotype, anterior facing right, ×2.3. (figured in Hisinger 1827, Pl. 7:7; 1828, Pl. 18:3). □H–L. *Colpomya hugini* n.sp. □H. Dorsal exterior view of articulated specimen, anterior facing left, note right valve being more convex than left, RMMo 25434, ×4.8. □I. Dorsal external view of articulated specimen, anterior facing right, note epibionts on shell, RMMo 25435, ×5.0. □J. External lateral view of left valve of articulated specimen, RMMo 25431, ×3.6. □K. External lateral view of left valve of articulated specimen, RMMo 25436, ×4.5. □L. External lateral view of right valve of articulated specimen, RMMo 25432, ×4.5. All specimens from the Wenlockian Mulde Beds at Djupvik, Gotland. (See Fig. 19 for explanation of muscular impressions.)

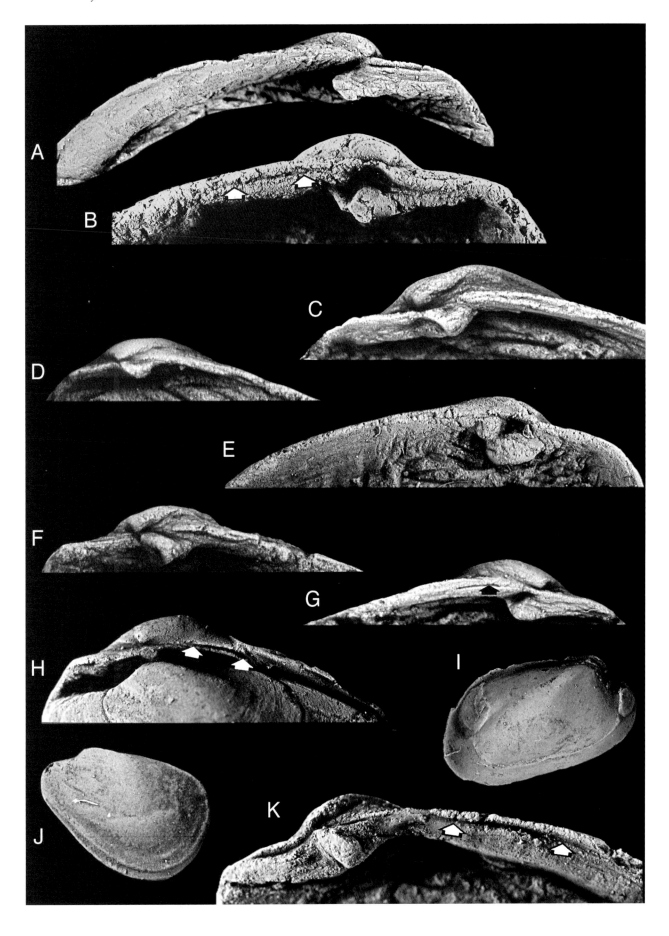

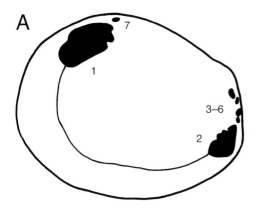

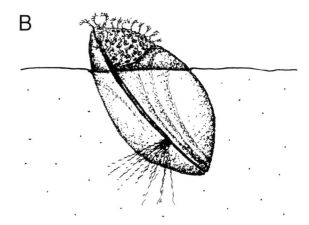

Fig. 40. Colpomya hugini n.sp. □A. Maximum number of muscular imprints in order of size. 1: Posterior adductor muscle scar. Note hypertrophied dorsoanterior part depicting byssal retractor muscle scars. 2: Anterior adductor muscle scar. 3–6: Pedal accessory muscle scars. 4: Pedal elevator muscle scar. 7: Anterior byssal–pedal retractor muscle scar. □B. Suggested life position.

convex, often almost straight; total length of shell of studied specimens 5–16 mm; L/H = 1.59; H/2W = 1.79.

Internal features. – Hinge line convex; dorsal margin entire below beak; ligament groove along dorsal margin (Fig. 39A–H, K); hinge plate strong, in right valve consisting of a single strong, triangular cardinal tooth in anterodorsal–posteroventral direction below and just posterior to beak (Fig. 39C–D, F, K); in left valve consisting of a deep triangular socket at the dorsal margin (Fig. 39A–B, E); posterior part of dorsal margin housing a conspicuous groove along and close to the shell margin, posterior to and originating from beak, extending to posterior part of dorsal margin; large, subquadrangular posterior adductor muscle scar with jagged anterior limitation, often with distinct growth lines (No.1 in Fig. 40A); extraordinarily deep subcircular anterior adductor muscle scar of moderate size, often with growth lines (No.2 in Fig. 40A); one deep, elliptical scar between anterior adductor muscle scar and umbonal cavity, close to the hinge line (No.3 in Fig. 40A); one deep, lanceolate scar in bottom of umbonal cavity (No.4 in Fig. 40A); one minute, circular scar close to and dorsal to the anterior adductor muscle scar (No.5 in Fig.

40A); 2–4 additional minute scars in umbonal cavity in an anterior position (Nos. 6–9 in Fig. 40A); pallial line nonsinuate, running from ventroposterior end of anterior adductor muscle scar to ventralmost end of posterior adductor muscle scar (Figs. 39I, 40A).

Remarks. – The shell shape of *Colpomya hugini* changes with size, the smallest specimens being less ovate (Fig. 37L) than the large ones (Fig. 37J). The length/height (L/H) ratio of 47 specimens ranges from 1.27 to 2.02 (average 1.59). In general outline the largest specimens of *C. hugini* exhibit a straight ventral margin while the smallest shells have a more rounded shape with a convex ventral margin.

In some shells the umbonal ridge is conspicuous, causing a flattening of the umbones and a slight sinus in the ventral margin (Fig. 37J).

Cross sections of articulated specimens show calcified remnants of the ligament (Fig. 17A), possibly being the inner ligament layer (see 'Morphology of the modiomorphid shell': cf. Owen *et al.* 1953). Although lacking ligament ridges, the ligament resembles that of *Mytilus edulis* (cf. Trueman 1950, pp. 227–228, Figs. 3–4).

The general arrangement of the muscular imprints of the present species is almost identical to that in e.g., *Modiodonta gothlandica* (cf. also similar muscular pattern of *Aleodonta burei* herein). It is also reminiscent of the principal muscular pattern of living Mytilacea. The hypertrophied posterior adductor muscle scar most probably is homologous with that of a posterior byssal–pedal retractor of extant mytilids, although not being as pronounced as in, e.g., the Recent *Modiolus modiola* and *Mytilus edulis* (cf Newell 1942, p. 30, Fig.6). Scar No. 3 in Fig. 40A of *Colpomya hugini* is heavily incised, elongate and conspicuously enlarged, compared with the remaining scars in the umbonal region. I suggest that it is the scar of a byssal retractor.

Fig. 39. Colpomya hugini n.sp. □A. Hinge of left valve. Note absence of lateral teeth, RMMo 25416, ×9.1. □B. Hinge of left valve, note ligament groove (arrows), RMMo 25418, ×12.5. C. Hinge of right valve, RMMo 25412, ×12.0. □D. Hinge of right valve, RMMo 25413, ×14.3. □E. Hinge of left valve, RMMo 25417, ×12.2. □F. Hinge of right valve, RMMo 25411 (holotype), ×15.3. □G. Hinge of left valve, note ligament groove (arrow), RMMo 25415, ×14.9. □H. Lateral view of dislocated articulated specimen showing insertion area for the ligament (arrows), anterior facing left, RMMo 25419, ×10.4. □I. Lateral view of internal cast of articulated specimen showing muscular impressions of right valve, RMMo 25420, ×3.7. □J. External lateral view of articulated specimen, RMMo 25433, ×10.0. □K. Hinge of right valve, note ligament groove (arrows), RMMo 25414, ×12.0. All specimens from the Wenlockian Mulde Beds at Djupvik, Gotland.

Comparisons. – *Colpomya hugini* seems to be most closely related to *C. munini*, from which it differs in having a more reinforced anterior part of the hinge plate, a thinner dorsal margin posterior to the beak, and a lower umbo. *C. hugini* also has more raised dorsal margins, less conspicuous diagonal sulcus, and a less pronounced ventral sinus than *C. munini*.

Colpomya hugini differs from ?*C. audae* as follows. In the right valve of *C. hugini* the entire hinge tooth is situated below and posterior to the beak, and the hinge is more reinforced in its anterior part, while in ?*C. audae* part of the hinge tooth is just below the beak. The socket of the left valve of *C. hugini* is deep and situated posterior to the beak, while in ?*C. audae* it is shallow and placed anterior to the beak. The length/height ratio of the most elongated specimen of *C. hugini* is 1.66, while in ?*C. audae* it ranges from 1.90 to 2.24, i.e. ?*C. audae* is characteristically more elongated than *C. hugini*.

Colpomya hugini differs from the type species of the genus, *C. constricta* Ulrich from the Ordovician of North America, in having a less pronounced tooth in the right valve and a less conspicuous protrusion in the anterior part of the hinge plate of the left one. The mesial depression is more pronounced in *C. constricta* than in *C. hugini*.

Colpomya demissa Ulrich 1894, p. 524, Pl. 36:21–22, has a pronounced triangular shell shape and a more conspicuous mesial sulcus and sinuated ventral margin than *C. hugini*. The internal features of *C. demissa* are unknown.

Modiolopsis nilssoni (Hisinger) in McCoy (1855, Pl. 1, I:21) resembles the most elongate specimens of *Colpomya hugini* at hand in gross shell shape but differs in exhibiting a pronounced umbonal ridge and a low umbo. Its interior is unknown.

Modiolopsis rigida Hall 1883 (illustrated in Hall 1885, Pl. 41:15) have similar external appearance as *Colpomya hugini*, but its interior is unknown. *Corallidomus versaillensis* (Miller, 1874), illustrated by Miller (1889, p. 491, Fig. 855), Pojeta (1971, Pl. 14:5) and Frey (unpublished thesis, Pl. 12:4–9), also shows similarities to *Colpomya hugini* in external features but has an edentulous hinge.

Gross shell shape of *Colpomya hugini* is similar to that of the edentulous *Aristerella nitidula* Ulrich (1897, p. 524, Pl. 35:30–39) except for its larger umbo.

Some species assigned to the edentulous *Modiolopsis* Hall show similarities in shell shape to *Colpomya hugini*, e.g., *Modiolopsis excellens* Ulrich (1897, p. 54, Pl. 36:13) and *Modiolopsis veterani* Barrande (1881, Pl. 259, III:20), but both these species have a slightly more rounded ventral margin than *Colpomya hugini*.

Modiolopsis leightoni Williams (1913, Pl. 29:7–10) is longer in relation to its height than is *Colpomya hugini*.

Modiolopsis parva Ulrich (1890, pp. 281–282) has inconspicuous beaks, is more compressed, has a straight hinge line, and an anterior margin that meets the dorsal margin at almost right angles.

The internal features of these species of *Modiolopsis* are unknown.

Ecology and habitat. – The shell of *Colpomya hugini* exhibits a number of external characteristics typical of both infaunal and epifaunal life habits, respectively. Its shell is streamlined, i.e. elongate, slightly inflated, with small beaks and a smooth shell surface, thus seemingly adapted either to active burrowing infaunal habitat or to epifaunal life in turbulent water.

However, since *Colpomya hugini* probably also had a byssus (see below), it was not likely an active burrrower. Furthermore, the inequivalved condition may indicate either low-angle fixation to the substrate (cf. Stanley 1972, p. 186, on the life habit of slightly similar, inequivalved pterineids) or a nestling, epibyssate life habit.

The diagonal umbonal ridge in, e.g., *Modiolus* and *Mytilus* channels the water to pass over the inhalent opening (posteroventrally) and over the exhalent opening (posterodorsally, Newell 1969, p. N:144, Fig. 88). The umbonal ridge of *Colpomya hugini* is better defined in the right valve. Thus, if the animal was fixed at a low angle with the left valve close to the substratum, the more developed umbonal ridge of the right valve may have channelled the water-flow more efficiently than would have been the case with the left valve up.

Colpomya hugini is also suggested to have had relatively strong byssal retractor muscles, and thus attached with byssal treads of corresponding strength. It seems as if this bivalve was resting on one valve in a low-angle, semi-infaunal way (Fig. 40B). It may also, however, have been fixed to sea-weed, coral stocks or similar structures. The juveniles, however, being more vulnerable to environmental forces and predators if situated above the sediment, had not yet achieved the straight ventral margin of the adult. Instead, in being rounded, the juvenile shell had a shell form better suited for burrowing. Accordingly, the juveniles may have lived within the sediment, in an infaunal position (cf. juveniles of *Modiolus americanus* who live entirely buried, while the adults have their posterior margin above the sediment surface, in Stanley 1972, p. 171; cf. also shell form of the earliest nepiontic stages of Recent *Avicula sterna* and *Perna epihippium*, both byssally attached when adult but lacking the conspicuous byssal notch when juvenile, in Jackson 1890, p. 328).

One well preserved, articulated specimen of *Colpomya hugini* has 13 *Spirorbis*-like tubes attached (6 on the right valve and 7 on the left, Fig. 37I). The presence of filter-feeding epibionts scattered all over both valves of an articulated bivalve specimen indicates: (1) that the epibionts were most likely attached while the bivalve was alive; and (2) that the bivalve individual was epifaunally attached to an object at some distance above the sediment surface (probably a floccular mud). As argued above, however, this species may also have been endobyssate, lying on one valve in a low angle fixation (reclining). In such a position, filter feeding epibionts may grow also on the valve facing the sediment (Savazzi 1984, p. 313, Text-fig. 1C).

Occurrence. – Wenlockian (Homerian) Mulde Beds at Djupvik, Gotland.

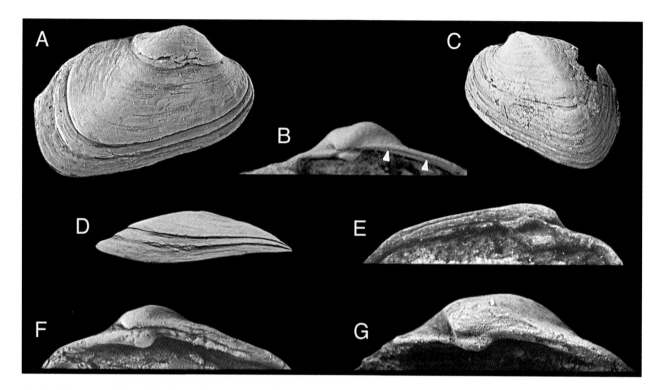

Fig. 41. Colpomya munini n.sp. □A. External lateral view of right valve, RMMo 781815, ×3.3. Valar 1. □B. Hinge of right valve (holotype). Note ligament groove (arrows), RMMo 16341, ×5.3. Burgsvik. □C. External lateral view of left valve, RMMo 152564, ×3.5. Vicinity of Burgsvik. □D. External ventral view of right valve (same as A). Note twisted ventral margin, ×4.5. □E. Hinge of left valve, RMMo 152466, ×4.7. Vicinity of Burgsvik. □F. Hinge of right valve, RMMo 791815, ×4.5. Burgsvik. □G. Hinge of right valve, LO 6291, ×11.2. Valar 1. All localities represent the Whitcliffian Burgsvik Beds of Gotland.

Colpomya munini n.sp.

Figs. 23:3, 25, 27, 41

Derivation of name. – After one of the old Scandinavian god Woden's two ravens, *Munin*, meaning Memory.

Holotype. – A right valve, RMMo 16341. Fig. 41B.

Type stratum. – Burgsvik Beds, Ludlovian.

Type locality. – Burgsvik, Gotland.

Material. – About 20 specimens.

Diagnosis. – Shell small, subequivalve, obliquely subovate, slightly compressed, twisted ventral margin, conspicuous umbonal ridge and sulcus, beaks medium-sized, hinge plate with a small cardinal tooth in right valve and shallow socket in left.

External features. – Shell small (maximum length ca. 18 mm), subovate (Fig. 41A, C), slightly inequivalved, right valve slightly more convex than left, obliquely subovate, inequilateral, compressed, margins even, except for twisted ventral margin (Fig. 41D), total length of shell about 1½ of total height; shell surface smooth with faint concentric growth lines and occasionally with irregular, conspicuous growth increment stops (Fig. 41A, C–D); beaks subanterior, small, close together, prosogyrate, not extending above sagittal plane; umbo well above dorsal margin; conspicuous umbonal ridge and mesial depression originating from umbo, which is more pronounced in right than in left valve; maximum convexity about mid-length and above mid-height of shell; no escutcheon; no lunule; dorsal margin long, convex; anterior margin short, narrow, almost pointed; posterior margin long, evenly rounded; ventral margin long, twisted; total length of shell in measured specimens 10.5–18.8 mm (a 14.9 mm); length/height = 1.48–1.88 (a 1.69); height/2×width = 1.38; number of specimens measured = 3.

Internal features. – Hinge line convex; hinge plate strong, in right valve containing a single, blunt tooth, slightly elongate in anterior–posterior direction below and immediately posterior to beak (Fig. 41B, F–G), a corresponding socket (Fig. 41E) in the left valve; no lateral element of hinge; dorsal margin holding one or two ligament grooves, one of which originating from the beak, and extending to posterior part of dorsal margin; no muscular impressions observed.

Remarks – Ontogenetic differences may be seen in the growth of the ligament area, which in juvenile specimens consists of one ligament groove and in adult ones of two grooves.

Comparisons. – *Colpomya munini* seems to be most closely related to *C. hugini* (see discussion of this species).

The Silurian North American *Modiolopsis exilis* Billings (1874, pp. 132–133, Pl. 8:5) closely resembles *Colpomya munini* in external features but differs in exhibiting a longer dorsal margin and in having a slightly crescent-shaped umbonal ridge situated more dorsally. The shell sculpture of *Modiolopsis exilis* consists of regular, strong concentric ridges. Internal features of this species are unknown.

Pulastra complanata Sowerby (1839, p. 609, Pl. 5:7) from the Silurian of Britain is also closely reminiscent of *Colpomya munini* but differs in having a lower umbo, a longer dorsal margin, and a crescent-shaped umbonal ridge. The internal features of the British species are unknown.

The Devonian North American *Modiomorpha regularis* Hall (1885, p. 270, Pl. 35:12) from the Shoharie Grit, superficially resembles *Colpomya munini* in external features but differs in having a smooth umbonal ridge and clearly prosogyrate beak, while the umbonal ridge of *C. munini* is obliquely straight and the beak more prosogyrate. The internal features of *Modiomorpha regularis* are unknown.

Modiomorpha tioga Hall (1885, p. 291, Pl. 40:18) and *Modiomorpha subalata* var. *chemungensis* Hall (1885, p. 284, Pl. 41:5–7), from the Devonian of North America, are both similar in shell shape to *C. munini* but their hinge morphology is unknown.

Ecology and habitat. – The specimens of *Colpomya munini* were found in oolitic limestone indicating a habitat with relatively high water energy and unstable, shifting bottom substrate. In Recent environments of this type, bivalves are either rapid burrowers in order to maintain a secure, buried life position, or they have thick shells which increases their stability (Stanley 1970, pp. 81–82). None of these characters are present in this species (cf. Fig. 23:3, where *C. munini* lies close to the region of slow burrowing).

The twisted commissure plane of this species suggests a semi-infaunal, byssally attached, life position with the commissure plane inclined, perhaps in a manner similar to that in some species of the Recent *Modiolus* (Fig. 22B; see discussion in 'Functional morphology'; cf. Savazzi 1984). The inequivalvity of *Colpomya munini* further supports the assumption of a low-angle life position. Probably, in this unstable environment, it was fixed with byssus treads to ensure a stable life position (Fig. 25).

Occurrence. – Ludlovian Hemse Beds at Linde klint (uncertain locality determination), Burgsvik Beds at Burgsvik, Bobbenarve, Fide, Rovalds, and Valar 1, Gotland.

?*Colpomya heimeri* n.sp.

Figs. 23:4, 25, 27, 42E, F

Derivation of name. – After the old Scandinavian name *Heimer*, meaning 'home'.

Holotype. – A left valve RMMo 152419. Length = 36.3 mm, height = 17.4 mm, width = 5.8 mm; L/H = 2.09, W/L = 0.16. Fig. 20E, F.

Type stratum. – Burgsvik Beds, Whitcliffian, Ludlovian.

Type locality. – Gansviken, Gotland.

Material. – About 10 specimens and fragments.

External features. – Shell medium-sized (maximum length ca. 46 mm), diagonally subrectangular (Fig. 42E), inequilateral, moderately compressed, posterior diagonal ridge extending from umbo to posteroventral extremity; margins even, shell surface with conspicuous growth-stop increments; beaks subanterior, prosogyrate, close together, slightly above dorsal margin; maximum convexity about mid-length and about mid-height of shell; no lunule; no escutcheon; dorsal margin slightly convex; anterior margin short, abruptly rounded; posterior margin long, rounded; ventral margin long, slightly concave, almost straight, in juvenile stages slightly convex; largest specimen about 46.5 mm long and 19.0 mm high.

Internal features. – Hinge line slightly convex, hinge plate of left valve with diagonal socket (Fig. 42F). No other features observed.

Comparisons. – See ?*Colpomya lokei* and ?*C. audae*.

Ecology and habitat. – This species has a shell shape suitable for rapid burrowing (Fig. 23:4). It shows similarities to some Recent endobyssate mytilids such as *Brachiodontes citrinus* and *Modiolus demissus* (Stanley 1972, p. 172, Text-fig. 4).

The shell of ?*Colpomya heimeri* is smooth and moderately thick. The species was found in a coarse oolitic limestone, indicating an unstable, shifting bottom substrate. In such a habitat, a thick shell may have been a hindrance to rapid burrowing, but if provided with a byssus (as suggested by the slightly concave ventral margin) the animal most probably could maintain a stable, semi-infaunal life position (Fig. 25).

Occurrence. – Ludlovian (Whitcliffian) Burgsvik Beds at Gansviken and Burgsvik, Gotland.

?*Colpomya vaki* n.sp.

Figs. 23:5, 27, 42A–D, G

Systematic position. – The hinge characters of ?*Colpomya vaki* are unknown, and therefore its systematic position can not be established, but its consistent shell form is strongly similar to that of *Colpomya*, and it is thus referred to this genus with a question mark.

Synonymy. – □?1892 *Modiolopsis similis* n.sp. – Ulrich, p. 237, Fig. 11. □?1895 *Modiomorpha westfalica* Beushausen – Beushausen, p. 26, Fig. 2.

Derivation of name. – After *Vak*, one of the many by-names of the old Scandinavian god Woden.

Fig. 42. □A–D. ?*Colpomya vaki* n.sp. □A. Internal casts of slightly opened articulated specimens *in situ*, RMMo 61123, ×1.5. Whitcliffian Klinta Formation at Ramsåsa, Scania. □B. External lateral view of fragmented left valve, RMMo 16295, ×2.4. Klinteberg Beds at Öjlemyrs kanal, Gotland. □C. Lateral view of left valve of articulated specimen, RMMo 61185, ×1.6. Whitcliffian Klinta Formation at Ramsåsa, Scania. □D. External lateral view of right valve (holotype), RMMo 16302, ×1.9. Klinteberg Beds at Öjlemyrskanal, Gotland. □E–F. ?*Colpomya heimeri* n.sp. □E. External lateral view of left valve (holotype), RMMo 152419, ×1.7. Whitcliffian Burgsvik Beds at Gansviken, Gotland. □F. Hinge of holotype (same as E), ×9.9. □G. ?*Colpomya vaki* n.sp. Lateral view of left valve of articulated specimen, RMMo 61202, ×1.8. Whitcliffian Klinta Formation at Ramsåsa, Scania.

Holotype. – A right valve, RMMo 16302, with shell preserved as recrystallized calcium carbonate. Fig. 42D.

Type stratum. – Klinteberg Beds, Homerian, Wenlockian.

Type locality. – Ölje Myrs kanal, Gotland.

Material. – About 20 specimens and fragments of disarticulated valves with preserved shells and ca. 30 articulated specimens preserved slightly opened as external moulds in a red, soft, silty and calcareous mudstone.

Diagnosis. – Shell medium-sized, diagonally subtriangular, moderately compressed, relatively large umbo, length/height ratio ca. 1.70.

Description. – Shell medium-sized (maximum length ca. 32 mm), diagonally subtriangular (Fig. 42C–D, G), inequilateral, moderately inflated, relatively large umbo, margins even, shell surface smooth with concentric growth lines, in some cases with growth-stop increments; beaks small, subanterior, prosogyrate, slightly above dorsal margin; no es-

cutcheon, no lunule; maximum convexity posterior to mid-length and slightly above mid-height of shell; dorsal margin slightly convex, almost straight; anterior margin short, strongly rounded; posterior margin long, slightly rounded, almost truncated; ventral margin long, rounded to almost straight; length 15.2–32.8 mm, height 8.2–18.8 mm, L/H = 1.65–1.85. No lateral elements of hinge; circular anterior adductor muscle scar: no other internal features observed.

Comparisons. – The Ordovician *Modilopsis similis* Ulrich, 1892, is similar to ?*Colpomya vaki* but is slightly more inflated, bears a more conspicuous mesial sulcus, and has a less pronounced umbonal ridge.

  ?*Colpomya vaki* shows some resemblance in external appearance to *Modiomorpha westfalica* Beushausen, 1895 but differs in exhibiting a straight ventral margin and a more pronounced prosogyrous beak.

Ecology and habitat. – No impressions of accessory muscles have been observed in ?*Colpomya vaki*, but the smooth, compressed shell suggests a rapid-burrowing ability (Fig. 23:5).

The species was found as disarticulated valves in coarse-grained sediments, and in the medium-size-grained Öved Sandstone as steinkerns of articulated specimens (Fig. 42A). Both these occurrences represent unstable habitats. This species shows great similarity to some Recent epibyssate mytilids, e.g., *Benthomodiolus abyssicola* (Knudsen, 1970; in Dell 1987, p. 32, Figs. 46–47) and *Idasola coppingeri* (Smith, 1885; in Dell 1987, p. 27, Figs. 13–16). Since there is no positive evidence of byssal attachment, and since in an unstable substrate rapid burrowing may be a means of maintaining a safe life position, ?*Colpomya vaki* is interpreted to have been non-byssate, and semi-infaunal.

Occurrence. – Wenlockian (Homerian) Klinteberg Beds at Ölje Myrs kanal, Gotland; Pridolian, Öved Sandstone Formation at Helvetesgraven and Ramsåsa, Scania.

?*Colpomya ranae* n.sp.
Figs. 23:7, 25, 27, 43

Derivation of name. – After *Ran*, a female sea giant of the old Scandinavian mythology.

Holotype. – A right valve, RMMo 158311, with preserved hinge. Fig. 43A–B.

Type stratum. – Burgsvik oolite, Burgsvik Beds, Ludlovian.

Type locality. – Locality unknown, Gotland.

Material. – Three specimens with preserved shell and 28 internal moulds (steinkerns).

Diagnosis. – Shell large, diagonally elliptical, equivalved, compressed, umbo and beaks inconspicuous, hinge with a large, robust cardinal tooth in right valve and socket in left valve, deep pedal muscle scars, shallow pallial sinus.

External features. – Shell large (maximum length ca. 70 mm), diagonally elliptical (Fig. 43E, G–H), equivalve (Fig. 43F), inequilateral, compressed, margins even, shell surface smooth with fine concentric growth lines; beaks subanterior, small (Fig. 43D), prosogyrate, close together, slightly above dorsal margin; posterior umbonal ridge extending to posteroventral extremity; maximum convexity at about mid-length and slightly above mid-height of shell; no escutcheon; no lunule; dorsal margin long, almost straight, slightly convex; anterior margin short, narrowly rounded, almost pointed; posterior margin long, evenly rounded; ventral margin long, slightly convex; total length of measured shells 45.4–70.5 mm (a = 48.94 mm; n = 9); L/H 1.65–1.91 (a = 1.78; n = 9); H/2W = 2.80–3.24 (a = 2.99; n = 6).

Internal features. – Hinge line convex, hinge plate with a single robust cardinal tooth in right valve below and posterior to beak (Fig. 43B); fitting into socket of left valve; no lateral teeth;

no ligament observed; large, faint, subelliptical posterior adductor muscle scar in a posterio-dorsal position, close to dorsal margin (Fig. 43G); medium-sized, deeply impressed subcircular anterior adductor muscle scar with jagged posterior limitation in a terminal position close to the anterior margin (Fig. 43C, G); one small, elliptical, extremely deeply impressed scar between anterior adductor muscle scar and umbo (Fig. 43C, F); One small, circular scar in bottom of umbonal cavity (Fig. 43F); two small scars in umbonal cavity in an anterolateral position (Fig. 43C); one minute scar in crest of umbonal cavity (Fig. 43F); 2–3 conspicuous ridges originating in the anterodorsal end of shell and running obliquely to posteroventral part, most pronounced in umbonal part (Fig. 43G–H); pallial line developed in many specimens as radial scars that run from ventralmost point of anterior adductor muscle scar, along, and at some distance from, ventral margin and meet ventralmost point of posterior adductor muscle scar in a shallow sinus-like feature (Fig. 43G–H).

Remarks. – The posterior adductor muscle scar shows no indication of hypertrophy, which is typical of byssate forms. The scar between the anterior adductor muscle scar and umbo is extremely deep and probably corresponds to an anterior pedal protractor, while the remaining scars probably show the positions of pedal elevators and pedal retractors. Early Palaeozoic bivalves generally lack a pallial sinus [with two exceptions, the Upper Ordovician *Lyrodesma*, in Kauffman 1969, p. N165, and the Upper Ordovician *Modiolopsis obtusas* in Pojeta 1985, p. 117) and probably had no siphons, or had only short ones. The pallial sinus of ?*Colpomya ranae* may indicate the presence of siphons.

The relatively deep impressions of pallial muscles of ?*Colpomya ranae* resemble the very strongly developed radiating pallial muscles in the Recent *Solemya velum* (in Drew 1900, p. 264). *Solemya velum* has united mantle margins where the radiating muscles are strongest. Therefore, it is not impossible that also ?*C. ranae* had fused mantle margins, facilitating active, rapid burrowing through the substrate and thus preventing sediment from entering the mantle cavity.

Comparisons. – ?*Colpomya ranae* differs from the other species of the genus in being conspicuously compressed, exhibiting a very low umbo, and having a significantly smaller beak.

Modiolopsis excellens Ulrich (1897, p. 511, Pl. 36:13–149, shows overall similarity to ?*Colpomya ranae* but differs from it in having a higher umbo and a less pronounced umbonal ridge. Its interior is unknown.

Ecology and habitat. – This species is particularly characterized by its equivalved, compressed, elongate shell, small beak, and smooth shell surface. It has conspicuous impressions of pedal muscles and bears strong impressions of radial muscles of the mantle (fused mantle margins?) but lacks positive evidence of byssal attachment. Shell morphology of ?*Colpomya ranae* suggests rapid burrowing (closest to blade shape in

Fig. 43. ?*Colpomya ranae* n.sp. □A. External lateral view of incomplete right valve (holotype), RMMo 58311, ×1.6. Burgsvik. □B. Hinge of the holotype, ×7.3. □C. Anterolateral view of anterior part of cast of articulated specimen showing pedal elevator scars (first two arrows from above), pedal retractor scar (third arrow), and anterior adductor scar (aa), RMMo 21565, ×2.3. Grötlingbo. □D. External lateral view of incomplete right valve, RMMo 158253, ×1.6. Gansviken. □E. Lateral view of internal impression of left valve showing anterior adductor muscle scar (aa), SGU Type 8370, ×1.6. Burgsvik. □F. Dorsal view of a cast of articulated specimen showing muscular impressions in anterior part (compare C), anterior facing left, RMMo 21546, ×1.6. Grötlingbo. □G. Lateral view of cast of articulated specimen showing scars of pedal elevators (first two arrows from the left), pedal retractor (third arrow), anterior adductor (aa), posterior adductor (pa), internal ridges, and pallial line (broad arrows), same specimens as in C, ×1.5. □H. Lateral view of internal mould of left valve showing anterior adductor muscle scar (aa), posterior adductor muscle scar (pa), internal ridges, pallial line (marginal arrows), and pallial sinus (arrow), RMMo 21544, ×1.0. Grötlingbo. All localities represent the Whitcliffian Burgsvik Beds of Gotland.

Fig. 23; cf. anatomy of rapid burrowing protobranch solemyids in Drew 1900 and Silurian protobranch *Janeia silurica* Liljedahl, 1984).

The species was found in argillaceous limestone, sandstone and oolitic limestone. The pallial sinus is indicative of si-

phons, and possibly this species was entirely infaunal with only the siphons protruding above the sediment (Fig. 25).

Occurrence. – Ludlovian (Whitcliffian) Burgsvik Beds at Burgsvik, Gansviken, Grötlingboudd, Gotland.

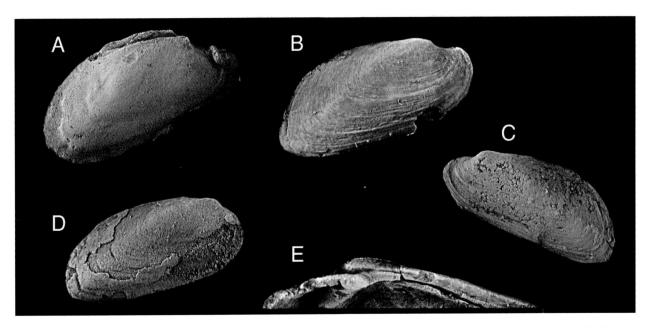

Fig. 44. ?*Colpomya audae* n.sp. □A. Lateral view of internal mould of right valve, SGU Type 8371, ×3.3. Burgsvik. □B. External lateral view of right valve, RMMo 16339, ×3.6. Burgsvik. □C. External lateral view of left valve (holotype), RMMo 16337, ×3.1. Burgsvik. □D. External lateral view of a right valve, LO 6345t, ×3.2. Valar 1. □E. Hinge of right valve, RMMo 21503, ×4.2. Burgsvik. All localities represent the Whitcliffian Burgsvik Beds, Gotland.

?*Colpomya audae* n.sp.

Figs. 23:9, 25, 27, 44

Derivation of name. – From the old Nordic name *Auda*, meaning swan-maiden, haze.

Holotype. – A complete left valve, RMMo 16337. Fig. 44C.

Type stratum. – Burgsvik Beds, Whitcliffian, Ludlovian.

Type locality. – Burgsvik, Gotland.

Material. – Fifteen specimens with preserved shell.

Diagnosis. – Shell medium-sized, obliquely elongated, beaks small, hinge plate with a cardinal tooth in right valve and a socket in left valve.

External features. – Shell medium-sized (maximum length ca. 53 mm), obliquely elongated (Fig. 44A–D), inequilateral, moderately inflated, margins even, shell surface smooth with concentric growth lines; beaks subanterior, small, close together, prosogyrate, not raised above sagittal plane, slightly above dorsal margin; umbonal ridge smoothly rounded; maximum convexity in anterior half and above mid-height of shell; no escutcheon, no lunule; dorsal margin long, slightly convex; anterior margin short, narrow and pointed; posterior margin long, rounded; ventral margin long, almost straight to slightly sinuated; total length of measured shell 14.70–53.40 mm (a = 29.33; n = 12); L/H 1.90–2.24 (a = 2.07; n = 12): H/2W = 1.84 (n = 1).

Internal features. – Hinge line convex, hinge plate narrow, in right valve with a single blunt cardinal tooth slightly elongate in anteroposterior direction, posterior and immediately be-low beak (Fig. 44E), in left valve with corresponding socket; no lateral teeth; posterior to and originating from beak a ligament groove along and close to shell margin extending about half way to posterior margin; large, faint, subelliptical posterior adductor muscle scar in a dorsoposterior position close to dorsal margin (Fig. 44A); medium-sized, subcircular, deeply impressed anterior adductor muscle scar with uneven posterior limitation close to anterior shell margin; one small, deeply incised scar between umbo and anterior adductor muscle scar; one small scar in bottom of umbonal cavity; pallial line non-sinuate, running from ventralmost point of anterior adductor muscle scar, at some distance from, and along ventral margin to ventralmost point of posterior adductor muscle scar.

Remarks. – The jagged posterior limitation of the anterior adductor muscle scar, and the deeply incised scar between the umbo and the anterior adductor muscle scar, may be the impressions of anterior pedal muscles, while the scar in the umbonal cavity may show the position of a pedal elevator.

Comparisons. – ?*Colpomya audae* shows similarities to ?*C. lokei* and ?*C. heimeri*. It differs from ?*C. lokei* in having a more elongated hinge tooth, a more pronounced diagonal sulcus and ventral sinus, and by its smaller size. Compared with ?*C. audae*, ?*C. heimeri* has a more distinctly angular lateral outline and conspicuous growth increment stops. ?*C. audae* exhibits external features similar to those of the Bohemian *Modiolopsis concors* Barrande (1881, Pl. 262 III:10–13), but since the internal morphology of the latter is unknown, the systematic relationships between the two forms can not be evaluated.

Ecology and habitat. – The size difference between the adductor muscle scars of ?*Colpomya audae* is not as pronounced as in extant endo-byssate mytilids (cf. Fig. 20A), suggesting intact burrowing ability. The overall shell characteristics, including a slight sinuation of the ventral margin, as well as the configuration of the muscle scars, suggest effective burrowing and a semi-infaunal, endobyssate life habit (Fig. 25).

This species was found in oolitic limestone indicating an original habitat of unstable, shifting environment. In such a habitat an elongate, thin and small-sized shell, such as that of the present species, would be advantageous for rapid burrowing, and along with the presence of a byssus, ensure a safe life position within the substrate.

Occurrence. – Ludlovian (Whitcliffian) Burgsvik Beds at Burgsvik, Glasskär, Rovalds, and Valar, Gotland.

?*Colpomya balderi* n.sp.

Figs. 23:6, 27, 45G–H, K

Derivation of name. – After *Balder*, the sun-god and son of Woden and Frigg, of the old Scandinavian mythology.

Synonymy. – □*v1880 *Modiolopsis Nilssoni* Hisinger – Lindström, in Angelin & Lindström, p. 59, Pl.19:5–6. □*non* 1880 *Ptychodesma Nilssoni* (Hisinger) – Lindström, in Angelin & Lindström, p. 18, Pl.2:21–22

Holotype. – Internal mould of an articulated specimen, RMMo 150416. Fig. 45G–H.

Type stratum. – Mulde Beds, Wenlockian.

Type locality. – Djupvik, Eksta Parish.

Material. – The holotype and one internal impression of a right valve.

Diagnosis. – Shell medium–sized, suboval, equivalve; beaks prosogyrate, close together, length of shell about 1.5 of height; strong hinge plate with one single robust cardinal tooth in the right valve, corresponding socket in left valve; deep anterior adductor muscle scar that is smaller than posterior one; ligament unknown.

External features. – Shell medium-sized (maximum length ca. 32 mm), suboval (45H, K), equivalve (Fig. 45G), inequilateral, moderately inflated, margins even, shell surface unknown, beaks subanterior, prosogyrate, maximum convexity in anterior and upper half of shell; total length of shell of studied specimens 30.2–31.5 mm; height 18.2–18.3 mm; width 4.5–4.75; L/H = 1.66–1.71; H/2W = 1.96.

Internal features. – Hinge line convex, hinge strong with a single robust cardinal tooth in right valve, corresponding socket in left valve (Fig. 45G); large, suboval posterior adductor muscle scar at some distance from posterior margin with tapering anterodorsal end (Fig. 45H), in one specimen with

distinct 'quick' and 'catch' structures (Fig. 19C): medium-sized, suboval anterior adductor muscle scar with jagged posterior limitation close to anterior margin, in one specimen with 'quick' and 'catch' structures (Fig. 45H); pallial line simple, running from ventroposterior end of anterior adductor muscle scar, at some distance from shell margin, to ventralmost end of posterior adductor muscle scar continuing from anterodorsal end to posterior adductor muscle scar along, and at some distance from, dorsal margin to hinge plate (Figs. 19C, 45H).

Remarks. – The jagged posterior part of the anterior adductor muscle scar probably indicates the presence of pedal–byssal accessory muscles (cf. Stanley 1970). The hypertrophied dorsal end of the posterior adductor muscle scar suggests that this species had strong pedal retractor muscles possibly also containing byssal retractors (cf. Newell 1942, p. 31).

Comparisons. – See *Modiodonta gothlandica*.

Ecology and habitat. – ?*Colpomya balderi* exhibits impressions of pedal muscles suggesting a functional foot and possibly also byssal attachment of the animal. In combination with shell shape, these features indicate a moderately rapid burrowing ability (Fig. 23:6) and a semi-infaunal, possibly byssally attached life position.

This species was found in a very argillaceaous calcilutite, which was once a carbonate mud mixed with terrigenous clay. In such a habitat an endobyssate bivalve may have been attached to larger particles such as shell fragments.

Occurrence. – Wenlockian (Homerian) Mulde Beds at Djupvik and Mulde tegelbruk, Gotland.

?*Colpomya lokei* n.sp.

Figs. 23:8, 27, 45A–E

Derivation of name. – After *Loke* in the old Scandinavian mythology. Loke was an evil–comic giant.

Holotype. – A fragmented right valve, RMMo 152715. Fig. 45A.

Type stratum. – Burgsvik Oolite, Burgsvik Beds, Whitcliffian, Ludlovian.

Type locality. – Burgsvik, Gotland.

Material. – Eight specimens preserved as internal moulds or as calcium carbonate shells.

Diagnosis. – Shell large, robust, subovally elongated, moderately inflated, beaks small, hinge plate with one robust cardinal tooth in right valve and socket in left.

External features. – Shell large (maximum length ca. 53 mm), robust, subovally elongated (Fig. 45B–C), inequilateral, faint ventral diagonal sulcus; margins even, shell surface smooth with concentric growth lines; beaks subanterior, prosogyrate,

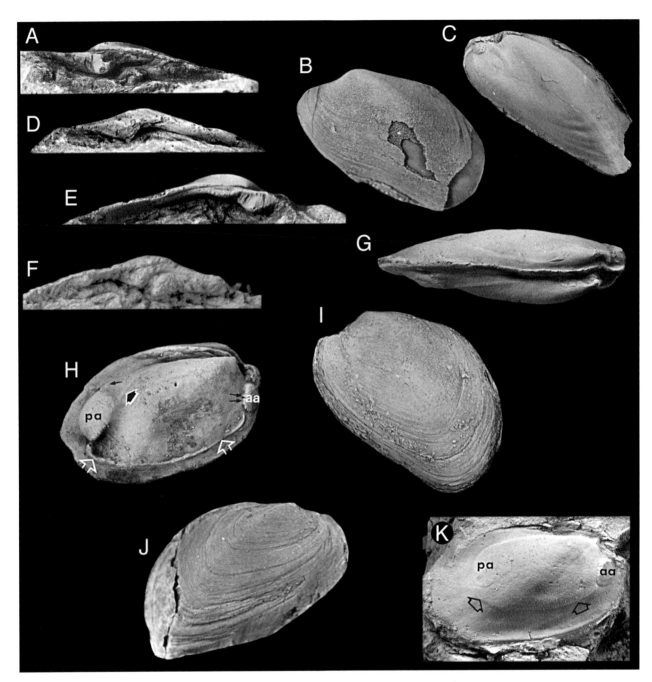

Fig. 45. □A–E. ?*Colpomya lokei* n.sp. □A. Hinge of right valve (holotype), RMMo 152715, ×2.4. Burgsvik, Burgsvik Beds, Gotland. □B. External lateral view of incomplete left valve, RMMo 152465, ×1.5. locality probably Burgsvik, Burgsvik Beds, Gotland. □C. Lateral view of internal mould of almost complete left valve, RMMo 18683, ×1.0. Burgsvik, Burgsvik Beds, Gotland. □D. Hinge of right valve, RMMo 21476, ×2.6. Burgsvik, Burgsvik Beds, Gotland. □E. Hinge of left valve (same as B), ×3.6. □F. ?*Colpomya friggi* n.sp. Hinge of left valve (holotype), RMMo 152469, ×3.5. Närsholm, Burgsvik Beds, Gotland. □G. ?*Colpomya balderi* n.sp. Dorsal view of internal cast of articulated specimen (holotype; figured by Angelin & Lindström 1880, Pl. 19:5–6), RMMo 150416, ×2.2. Djupvik, Mulde Beds, Gotland. □H. ?*Colpomya balderi.* Lateral view of internal mould of right valve of articulated specimen (holotype, same as G, figured by Angelin & Lindström 1880, Pl. 19:5), showing impressions of posterior adductor muscle (pa), anterior adductor muscle (aa), posterior pedal–byssal retractor (left black arrow), anterior pedal accessory muscles (right black arrows), secondary line of pallial attachment (black & white arrow) and pallial line (broad white arrows), see enlargement in Fig. 19C, ×1.7. □I. ?*Colpomya friggi.* external lateral view of left valve (holotype, same as F), ×1.0. □J. ?*Colpomya* sp. 1. External lateral view of right valve, LO 6292t, ×1.2. Glasskär, Burgsvik Beds, Gotland. □K. ?*Colpomya balderi.* Lateral view of internal mould of right valve showing posterior adductor muscle scar (pa), anterior adductor muscle scar (aa), and pallial line (arrows), RMMo 21708, ×1.6. Djupvik, Mulde Beds, Gotland.

small, close together, somewhat above dorsal margin; maximum convexity about mid-length and at mid-height of shell; no escutcheon; no lunule; dorsal margin long, slightly convex; anterior margin somewhat rounded; posterior margin evenly rounded; ventral margin long, slightly convex, almost straight; length of measured specimens 41.6–53.3 mm, height 21.5– 24.1 mm, width 7.40–7.25 mm; L/H = 1.93–2.23, H/2W = 1.51 ($n = 2$).

Internal features. – Hinge line slightly convex, hinge plate in right valve with a robust, diagonal cardinal tooth, below and slightly posterior to beak (Fig. 45A, D), in left valve with a socket, anterior to which the ligamental plate is strong (Fig. 45E); longitudinal ligament furrow in dorsal margin posterior to beak (Fig. 45E), small, subcircular anterior adductor muscle scar close to anterior margin (Fig. 45C); posterior adductor muscle scar not observed; pallial line close to ventral margin.

Comparisons. – ?*Colpomya lokei* seems to be most closely related to ?*C. heimeri* from which it differs in the following way. Generally the anterior lobe of ?*C. lokei* is somewhat more pronounced than that of ?*C. heimeri*, which has a more distinct diagonal ridge. The shell of ?*C. lokei* is smooth, but in ?*C. heimeri* conspicuous growth increment stops are present.

Ecology and habitat. – The elongated shape (including a convex ventral margin), the smooth shell surface and rather thin shell of ?*Colpomya lokei* are features indicative of relatively rapid burrowing (Fig. 23:8). No muscular impressions, except for the anterior adductor muscle scar, have been observed. Thus, no positive evidence of byssal attachment is at hand. Probable life position is with the posterior end just above the sediment surface.

?*Colpomya lokei* is found in oolitic limestone. In such a coarse-grained sediment there is no need for having the inhalent and posterior openings high above the sediment surface as must be the case in fine-grained sediments where there is risk of fouling near the sediment surface.

Occurrence. – Ludlovian (Whitcliffian) Burgsvik Beds at Burgsvik and Snäckvik, Gotland.

?*Colpomya friggi* n.sp.
Figs. 23:10, 25, 27, 45F, I

Derivation of name. – After the old Scandinavian mythological sea-giant *Frigg*.

Holotype. – A complete left valve, RMMo 152469. Fig. 45F, I.

Material. – The holotype and additional fragments preserved with calcium carbonate shells.

Type stratum. – Burgsvik Beds, Ludlovian.

Type locality. – Glasskär, Gotland.

Diagnosis. – Shell large, robust, erect, diagonally subtriangular, conspicuous umbonal ridge, umbo and beaks small, hinge with a robust tooth in righ valve and a socket in left valve.

Description. – Shell large (maximum length ca. 57 mm), diagonally subtriangular (Fig. 45I), inequilateral, moderately inflated, margins even, shell surface smooth with concentric growth lines; beaks small, subanterior, medium-sized, close together, prosogyrate, hardly reaching above dorsal margin; umbonal ridge dividing shell in two equal parts; maximum convexity slightly anterior to mid-length and in upper half of shell; no escutcheon; no lunule; dorsal margin convex; anterior margin short, convex; posterior margin long, evenly rounded; ventral margin long, convex in anterior part and slightly sinuated in posterior part; total length of measured shell 56.8 mm; height 39.1 mm; width 11.75 mm; L/H = 1.45; H/2W = 1.67. Hinge line convex, hinge plate strong with a robust cardinal tooth in the right valve (Fig. 45F); left valve with a diagonal socket just below beak.

Comparisons. – In lateral outline ?*Colpomya friggi* resembles ?*C.* sp. 1 and ?*C. vaki* but it is larger and more inflated than these and has a more distinct umbonal ridge and less pronounced anterior lobe than ?*C. vaki*. Furthermore, the ridge of ?*C. friggi* is more diagonally set and the shell more erect than in ?*C.* sp. 1.

Ecology and habitat. – Based on shell shape alone it seems as if ?*Colpomya friggi* was a slow-burrowing bivalve (Fig. 23:10). No muscular impressions have been observed. Possibly the slightly sinuated part of the ventral margin indicates byssal attachment.

The sediment in which this species was found is an oolitic limestone, suggesting an unstable original environment. It possibly lived semi-infaunally attached with a byssus (Fig. 25).

Occurrence. – Ludlovian (Whitcliffian) Burgsvik Beds at Glasskär, Gotland.

?*Colpomya* sp. 1
Fig. 23:11, 27, 45J

Systematic position. – The hinge of ?*Colpomya* sp. 1 has not been observed, but its similarity in gross shell shape and shell thickness to ?*C. friggi* possibly indicates close systematic relationship.

Material. – Three specimens and additional fragments with preserved shell.

Description. – Shell large (maximum length ca. 51 mm), robust, diagonally subtriangular (Fig. 45J), inequilateral, moderately inflated, margins even, shell surface smooth with concentric growth lines; beaks small, subanterior, medium-sized, close together, prosogyrate; umbonal ridge prominent, closer to posterior margin; inconspicuous mesial depression; maximum convexity at about mid-length and in upper half of shell; no escutcheon; no lunule; dorsal margin short, convex; anterior margin short, rounded; posterior margin long, evenly rounded; ventral margin long, slightly convex; L = 51.30 mm; H = 32.5 mm; W = 8.8 mm; L/H = 1.58; H/2W = 1.85.

Comparisons. – Externally ?*Colpomya* sp. 1 agrees with the Middle Devonian *Modiomorpha mytiloides* (Conrad) as figured by McAlester (1962, Pl. 16:10; composite mould lacking hinge structures).

?*Colpomya* sp. 1 seems to be most closely related to ?*C. friggi*, from which it is distinguished by its somewhat different shell shape, a more compressed shell, and a more distinct umbonal ridge situated less dorsally than in ?*C.* sp. 1.

Ecology and habitat. – Shell morphology of ?*Colpomya* sp. 1 suggests a moderately rapid-burrowing ability (Fig. 23:11). No muscular impressions have been observed, but the shell form, with moderately reduced anterior end, indicates a semi-infaunal life position with the posterior end protruding above the sediment.

This species was found in a medium-grained oolitic limestone indicating an unstable environment. There is no positive evidence of a byssus, rather the contrary (convex ventral margin), and in this kind of environment a moderately rapid burrowing ability probably would have been sufficient to maintain a safe infaunal life position.

Occurrence. – Ludlovian (Whitcliffian) Burgsvik Beds at Glasskär, Gotland.

Genus *Aleodonta* n.gen.

Figs. 16, 36

Derivation of name. – After the old Scandinavian name *Ale* and Greek *odous* (genitive *odontos*), tooth.

Type and only species. – *Aleodonta burei* n.sp.

Diagnosis. – Shell small, subequivalve, inequilateral, subovate, margins even except for dorsal margin of right valve overlapping left; beaks subanterior, small, prosogyrate; no diagonal sulcus; shell surface smooth; no escutcheon; lanceolate lunule; internal linear ligament, dorsal margin of left valve sigmoidal below beak; hinge line convex; no hinge plate; one cardinal tooth in the right valve, one socket in the left; no lateral hinge teeth; anterior adductor muscle scar deep, smaller than posterior one; a number of accessory muscle scars, integripalliate.

Comparisons. – *Aleodonta* seems to be most closely related to *Modiodonta*, *Colpomya*, and *Mimerodonta*. It differs from all three in the overlapping posterior part of the dorsal margin, the sigmoidal dorsal margin of the left valve below the beak, and from *Modiodonta* and *Colpomya* in lacking an ordinary

hinge plate. *Aleodonta* differs from *Mimerodonta* by exhibiting one cardinal tooth in the right valve and one socket in the left. *Mimerodonta* has one cardinal tooth flanked by two shallow sockets in the right valve, and one deep socket flanked by a tooth-like reinforcement on each side in the left valve.

Aleodonta burei n.sp.

Figs. 16B, 17B, 19D, E, 23:12, 27, 46–47

Derivation of name. – After *Bure*, the progenitor of the gods in the old Scandinavian mythology.

Synonymy. – □1921 *Modiolopsis nilssoni* (Hisinger) – Hede, pp. 51,55. □1927 *Modiolopsis nilssoni* (Hisinger) – Hede (in Munthe *et al.*), pp. 20, 21, 54.

Holotypes. – One complete right valve RMMo 25421, Fig. 46M and a complete left valve RMMo 25424, Fig. 46A.

Type stratum. – Mulde Beds, Homerian, Wenlockian.

Type locality. – Djupvik 1–2, Gotland, Sweden. (see Jaanusson 1986 for the extent of this composite locality)

Material. – About 300 specimens, most of which are articulated and with preserved shell.

Diagnosis. – Shell small, obliquely subovate, slightly inequivalve, in most cases right valve more convex than left one which it overlaps, moderately inflated, undefined umbonal ridge, lanceolate lunule; hinge with a robust cardinal tooth in right valve, and a deep socket in left valve.

External features. – Shell small (maximum length ca. 17 mm), obliquely subovate (Fig. 46B–D, I, K), inequilateral, moderately inflated (Fig. 46C), margins even except for dorsal margin of right valve slightly overlapping left (Fig. 46F–G); total length of shell about 1.3 of total height; shell surface

---

Fig. 46. Aleodonta burei n.sp. □A. Hinge of left valve, RMMo 25424, ×11.0. □B. External lateral view of left valve of articulated specimen, RMMo 25428, ×4.7. □C. Dorsal external view of articulated specimen, anterior facing left, RMMo 25426, ×5.5. □D. External lateral view of left valve of articulated specimen, same as C, ×4.7. □E. Hinge of left valve, RMMo 25423, ×12.0. □F. Anterior external view of articulated specimen, same as C and D, ×4.2. □G. External ventral view of articulated specimen. Note epibionts in posterior part of right valve (compare J), RMMo 25427, ×3.5. □H. Hinge of left valve, RMMo 25425, ×8.6. □I. Lateral view of left valve of internal cast of articulated specimen showing muscular impressions of anterior adductor muscle (aa), anterior accessory muscles (as), posterior pedal retractor muscle (pbr), posterior adductor muscle (pa), and pallial line (arrows). RMMo 25428, ×4.5. □J. External lateral view of right valve of articulated specimen. Note epibiont colony at posterior end (same as G), ×3.5. □K. Lateral view of internal mould of left valve of articulated specimen showing muscular impressions of anterior adductor muscle (aa), and posterior adductor muscle (pa) RMMo 25430, ×4.2. □L. Hinge of right valve, RMMo 25422, ×11.0. □M. Hinge of right valve (holotype), RMMo 25421, ×15.0. All specimens from the Homerian Mulde Beds at Djupvik, Gotland.

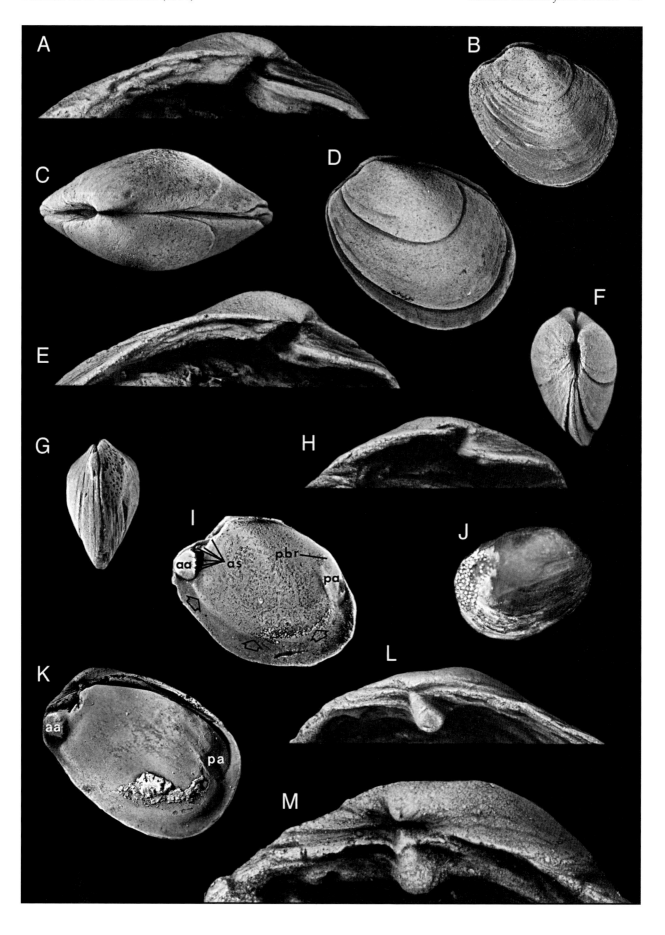

smooth with conspicuous irregular growth rugae; beaks sub-anterior, small, close together, prosogyrate, located slightly above dorsal margin, umbonal ridge rounded, undefined; maximum convexity in anterior half and above mid-length of shell; no escutcheon, lanceolate lunule; dorsal margin long, convex to strongly convex; anterior margin short, narrow, and abruptly rounded; posterior margin evenly rounded, posterior end of this margin almost truncated in some specimens; ventral margin rounded; total length of shell of measured specimens 7–17 mm; length/height = 1.34; length/width = 2.35; height/width = 1.67.

Internal features. – Hinge line convex, hinge 'support' strong, containing one cardinal tooth below and just posterior to beak of the right valve (Fig. 46L–M), that fits into socket of the left (Fig. 46A, E, H); no lateral element of hinge; posterior to the beaks the dorsal margin is slightly incised to hold linear ligament; faint, subquadrangular, large posterior adductor muscle scar (Fig. 46I, K), sometimes with an irregular, jagged anterior end and growth lines as well as 'quick' and 'catch' portions well marked (No. 1 in Fig. 47A); extremely deep, subelliptical anterior adductor muscle scar, in some specimens with growth lines and 'quick' and 'catch' structures (No. 2 in Fig. 47A); three to four accessory muscle scars between anterior adductor muscle scar and the bottom of the anterior part of umbonal cavity in anterodorsal position (Nos. 3–6 in Fig. 47A); one minute, circular scar anterodorsal to posterior adductor muscle scar (No.7 in Fig. 47A); pallial line non-sinuate, running from posteriormost end of anterior adductor muscle scar, at some distance from the shell margin, to ventral point of posterior adductor muscle scar.

Remarks. – The length/height ratio of 27 articulated specimens ranges from 1.14 to 1.50.

Cross sections of articulated specimens exhibit calcified remnants of the ligament (probably the inner layer of the ligament, Fig. 17B. cf. Fig. 18), similar to that of *Mytilus* edulis, except for the absence of ligamental ridges (cf. Trueman 1950, pp. 227–228, Figs. 3–4).

The general pattern of muscular impressions of *Aleodonta burei* is similar to that in fossil and extant Mytilacea (cf. also *Colpomya hugini* and *Modiodonta gothlandica*). The irregular limitation of the anterior end of the posterior adductor muscle scar (Figs. 25A, 46K) may be homologous with a posterior byssal and pedal retractor in modern mytilaceans, although it is not as large as in the extant *Modiolus modiolus* and *Mytilus edulis* (pbr in Fig. 20). The insertions of the accessory muscles in the anterior portion of the umbonal cavity of *Aleodonta burei* have a corresponding position as the anterior byssal retractors of the mentioned extant mytilaceans and, thus, the corresponding muscles probably had the same or similar function (cf. Newell 1942, Fig.6).

Ecology and habitat. – *Aleodonta burei* has a rounded, gibbous shell with a smooth surface, thus resembling some burrowing

species, such as representatives of the ovate Nuculoida (cf. Liljedahl 1983, 1984). The height/width plotted against length/height gives the value 1.25, thus a figure that falls within the region of slow burrowing but close to the rapid burrowing region in Fig. 23. The slightly inequivalve nature of the valves, however, suggests that the species was not an active burrower but buried to get a semi-infaunal life position [cf. *Yoldia* and other protobranchs (infaunal deposit feeders) that sometimes assume a semi-infaunal position; see Yonge 1941].

The Djupvik marl was once a soft carbonate mud mixed with terrigenous clay. To avoid silting of the gills, the filter-feeding epifauna must live at some distance above the sediment surface in such a habitat. Generally in a mud, the only attachment surfaces available for an epifauna are furnished by other organisms. In deep water muds, mainly molluscan skeletal remains provide a substratum for epifauna (Allen 1953).

Eight articulated specimens of *Aleodonta burei* contain epibionts on their right valves (Fig. 46G, J). The epibionts are bryozoans, which in all cases but one leave the valve margin free so that the valves of the bivalve could operate. In the remaining specimen, the bryozoan colony has slightly overgrown the right valve, seemingly having made it impossible for the bivalve to function. However, specimens of *Spirorbis* may extend over the sagittal plane attached to both valves of live bivalves (P.A. Berkman, Ohio State University, personal communication, 1991)

Since all specimens with attached epibionts are articulated, they must have been buried rapidly after death. The epibionts do not cover large areas of the bivalve shell but are restricted to the posterior end. Thus, the epibionts, being filterers, evidently were attached to live bivalves.

In analogy with Recent examples of epifauna on bivalves (Savazzi 1984) the position of the epibionts of *Aleodonta burei* suggest a life position with the posterior end of the shell protruding above the surface of the bottom (Fig. 47B). On the single specimen with the epibiont slightly overgrowing the valve margins, the epibiont possibly kept growing for some time after the bivalve's death but eventually became buried together with the dead bivalve. The fact that all the epibionts are situated on the right valve of all infested specimens may indicate that this species had a low-angle position in the sediment (Fig. 25B; cf. Stanley 1972, p. 185).

The arrangement of the muscular imprints suggests byssal attachment. Compared with, e.g., *Colpomya hugini* and *Modiodonta gothlandica*, the byssal attachment of *Aleodonta burei* was evidently weaker, as indicated particularly by the smaller size of the byssal retractor muscle scar. This condition may have been compensated by the low-angle life position with the commissure plane more horisontal.

Occurrence. – Wenlockian (Homerian) Mulde Beds at Djupvik, Gotland.

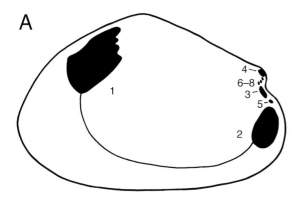

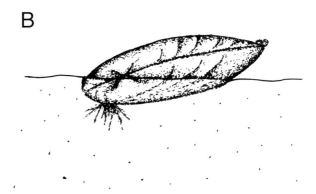

Fig. 47. Aleodonta burei n.sp. □A. Maximum number of muscular imprints in order of size. 1: Posterior adductor muscle scar. Note jagged dorsoanterior part indicating accessory muscle scars. 2: Anterior adductor muscle scar also with one side jagged, indicating the presence of pedal accessory muscles. 3–6: Anterior byssal–pedal accessory muscle scars. 7: Posterior pedal–byssal retractor muscle scar. B: Suggested life position. Note bryozoan epizoan colony in posterior part of right valve.

Genus *Mimerodonta* n.gen.

Figs. 16, 36

Derivation of name. – After the old Scandinavian god *Mimer* and Greek *odous* (genitive *odontos*), tooth.

Type species. – *Mimerodonta atlei* n.sp.

Species. – *Mimerodonta atlei, Mimerodonta njordi.*

Diagnosis. – Shell small, inequilateral, diagonally subtriangular to elongate, margins even, beaks subanterior, prosogyrate, diagonal sulcus, shell surface smooth; no escutcheon, lanceolate lunule; internal linear ligament; dorsal margin continuous below beak; hinge line almost straight; no hinge plate; one robust cardinal tooth with one to two shallow sockets on each side in the right valve, one deep socket flanked by two tooth-like reinforcements in the left; no lateral elements of the hinge; anterior adductor muscle scar deep, smaller than posterior one; a number of accessory muscle scars; integripalliate.

Comparisons. – See discussion of *Modiodonta.*

Mimerodonta atlei n.sp.

Figs. 16C, 23:13, 27, 48, 49

Derivation of name. – After the old Scandinavian mythological king *Atle*, meaning 'little father'.

Holotype. – A right valve, RMMo 24094. Fig. 48D.

Type stratum. – Ludlovian Hemse Beds.

Type locality. – Bjärs Träskbacke, Gotland.

Material. – About twenty specimens and fragments.

Diagnosis. – Shell small, obliquely subtriangular, moderately inflated, diagonal umbonal ridge, lanceolate lunule, hinge with a robust triangular, cardinal tooth flanked by two shal-

low sockets in right valve and a deep socket with projecting tooth-like reinforcements of the dorsal margin on each side in left valve.

External features. – Shell medium-sized (maximum length ca. 17 mm), obliquely subovate to subtriangular (Fig. 48A, E–G), inequilateral, total length of shell almost $\frac{1}{2}$ of total height; shell surface smooth with fine concentric growth lines, in some cases with distinct growth increment stops; beaks subanterior, small, close together, prosogyrate, not extending above sagittal plane, slightly above dorsal margin; mesial depression; maximum convexity in anterior half of shell, and slightly below mid-height; no escutcheon; lanceolate lunule; dorsal margin long, convex; anterior margin short, rounded; posterior margin long, evenly rounded; ventral margin long, slightly sinuated; total length of shell of measured specimens 14.1–17.4 mm; L/H = 1.24–1.38 ($a = 1.30$; $n = 6$); H/2W = 1.32–1.50 ($a = 1.40$; $n = 3$).

Internal features. – Hinge line convex; hinge in the right valve with a single robust, blunt cardinal tooth immediately below beak, flanked by a shallow socket-like depression on each side (Fig. 48B–D). Left valve has deep, V-shaped socket with tooth-like projecting dorsal margins on each side (Fig. 48C, H); posterior part of dorsal margin broad, containing narrow ligament grooves originating from the beak, that reach posterior part of dorsal margin; extremely large subquadrangular posterior adductor muscle scar in dorso-posterior position, close to dorsal margin; deep, subrounded anterior adductor muscle scar; one small, deep scar between anterior adductor muscle scar and umbonal cavity in a dorsal position; pallial line simple, non-sinuate.

Comparisons. – *Mimerodonta atlei* differs from *M. njordi* in being subtriangular and in exhibiting a distinct umbonal ridge. *M. njordi* is elongate and has a less distinct umbonal ridge.

Mimerodonta atlei shows some resemblance to the Ordovician *Modiomorpha kallholniensis* Isberg (1934, p. 171, Pl.

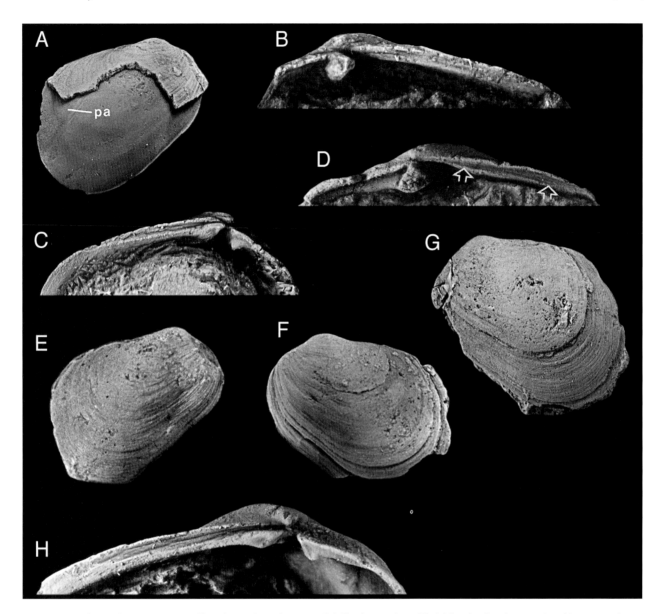

Fig. 48. Mimerodonta atlei n.sp. □A. External lateral view of a partly preserved shell and internal mould of right valve showing posterior adductor muscle scar (pa), RMMo 152543, ×3.3. □B. Hinge of right valve, RMMo 152543, ×6.7. □C. Hinge of left valve, note ligamental groove (at arrows), RMMo 152242, ×4.2. □D. Hinge of right valve (holotype), RMMo 24094, ×8.7. □E. External lateral view of right valve, RMMo 158262, ×5.7. □F. External lateral view of left valve, RMMo 152541, ×3.5. □G. External lateral view of left valve, RMMo 152542, ×3.3. □H. Hinge of left valve, RMMo 152541, ×8.3. All specimens from the Ludlovian Hemse Beds at Bjärsträskbacke, except for E which was collected at Ljugarn, all Gotland.

26:5). Externally *Modiomorpha kallholniensis* is more elongate, has a less pronounced posterior end, and a more conspicuous mesial sulcus. The dentition of the Ordovician species is faint, consisting of a diminutive cardinal tooth in the right valve (cf. robust cardinal tooth of *Mimerodonta atlei*).

Mimerodonta atlei also resembles the Bohemian *Modiolopsis interpolata* Barrande (1881, Pl. 258, I:5,3), the interior of which, however, is unknown.

Ecology and habitat. – The subtriangular and gibbous shell morphology of *Mimerodonta atlei* suggests a slow-burrowing ability (Fig. 23:13). The deep anterior accessory muscle scar indicates either a functional foot or a strong byssal retractor. The presence of an anterior lobe and the fact that the centre of gravity is relatively high, suggest a semi-infaunal life position, possibly with byssal attachment (Fig. 49).

The species was found in a dense, fine-grained limestone, which was once a soft calcareous mud.

Occurrence. – Ludlovian Hemse Beds at Bjärs Träskbacke, Hageby Träskbacke, and Ljugarn, Gotland.

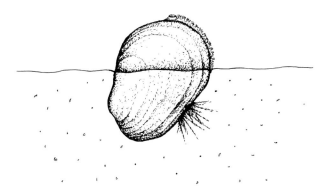

Fig. 49. Suggested semi-infaunal life position of *Mimerodonta atlei* n.sp.

Mimerodonta njordi n.sp.

Figs. 23:14, 25, 27, 50

Derivation of name. – After *Njord*, god of wealth, trading and shipping, in the old Scandinavian mythology.

Holotype. – A left valve, RMMo 16335. Fig. 50C.

Type stratum. – Burgsvik Beds, Whitcliffian, Ludlovian.

Type locality. – Burgsvik, Gotland.

Material. – About a hundred specimens.

Diagnosis. – Shell medium-sized, obliquely elongate, sub-equivalve with right valve slightly more convex than left, beaks small, hinge plate with one robust cardinal tooth in right valve and deep socket in left valve, hinge line straight posterior to beaks, ventral margin slightly sinuated.

External features. – Shell medium-sized (maximum length ca. 20 mm), obliquely subovate to elongate (Fig. 50A–C, E, H), subequivalved, right valve slightly more convex than left; inequilateral, compressed, margins even, total lenght of shell about $1\frac{1}{3}$ of total height; shell surface smooth with fine concentric growth lines, sometimes with irregular conspicuous growth-increment stops; beaks subanterior, small, close together, extending slightly beyond sagittal plane, moderately raised above dorsal margin; diagonal umbonal ridge evenly rounded; maximum convexity in anterior half and above mid-height of shell; no escutcheon; no lunule; dorsal margin long, slightly convex, almost straight posterior to beaks; anterior margin short, evenly rounded, almost truncated; posterior margin evenly rounded; ventral margin long, slightly sinuated; L/H = 1.53–2.22 (a = 1.75); H/2W = 1.09–1.47 (a = 1.32); n = 5.

Internal features. – Hinge line straight; hinge in right valve with a single, blunt, slightly elongate cardinal tooth arranged in dorso-anterior/ventro-posterior direction below and just posterior to beak, flanked by shallow sockets on each side (Fig. 50D, F, K); left valve containing a corresponding deep socket flanked by projecting reinforcements on each side (Fig. 50G,

I–J); no lateral teeth; posterior part of dorsal margin housing a conspicuous broadening of the dorsal margin with ligament grooves along, and at some distance from, the valve margin and originating from beak, reaching the posterior part of the dorsal margin; large, faint, quadrangular posterior adductor muscle scar in a posterior position, close to the dorsal margin (Fig. 50H); subequal, relatively deeply incised, subcircular, terminal anterior adductor muscle scar close to the anterior margin; one circular, deeply impressed scar between anterior adductor muscle scar and beak in a dorsoanterior position; pallial line simple, non-sinuate, running from ventroposterior end of anterior adductor muscle scar to ventralmost end of posterior adductor muscle scar (Fig. 50H).

Remarks. – Although no articulated specimens have been found, it appears that the right valve of this species is slightly more convex than the left one.

Comparisons. – See *Mimerodonta atlei. Mimerodonta njordi* shows similarity in external features to *Modiolopsis interpolata* Barrande (1881, Pl. 258, I:8; *non Modiolopsis interpolata* Barrande, 1881, Pl. 258, I:5,3) but differs in having a slightly sinuated ventral margin and a smaller beak.

Ecology and habitat. – Shell width of *Mimerodonta njordi* is relatively large in relation to height resulting in a position close to the region of slow-burrowing in the diagram in Fig. 23:14. The presence of a deep accessory muscle scar in the anterior region, and the slightly sinuated ventral margin, indicate byssal attachment.

These specimens were isolated from a coarse oolithic limestone, indicating an unstable, shifting original environment. A moderately thick shell attached with byssus most probably compensated for a slow burrowing ability in such a habitat, and the species may have had a semi-infaunal life position (Fig. 13; cf. extant *Modiolus demissus,* see Stanley 1972, p. 172, Text-fig. 4B).

Occurrence. – Ludlovian (Whitcliffian) Burgsvik Beds at Burgsvik, Fide, Grötlingboudd (uncertain determination of species), Rovalds, and Valar, Gotland.

Genus *Radiatodonta* Dahmer, 1921

Figs. 16, 36

Type species. – *Radiatodonta goslarensis* Dahmer 1921, p. 245, Pl. 10:6–7.

Diagnosis. – Modiomorphid with several short, erect, dorsally convergent hinge teeth.

Comparisons. – This genus is easily distinguished from other modiomorphids by its conspicous arrangement of several convergent hinge teeth. In the type species, *Radiatodonta goslarensis* from the Devonian of Germany, the teeth are dorsally convergent, but Zhang (1977, Pl. 120:17) described the Devonian *Radiatodonta shaodongensis* Zhang from China

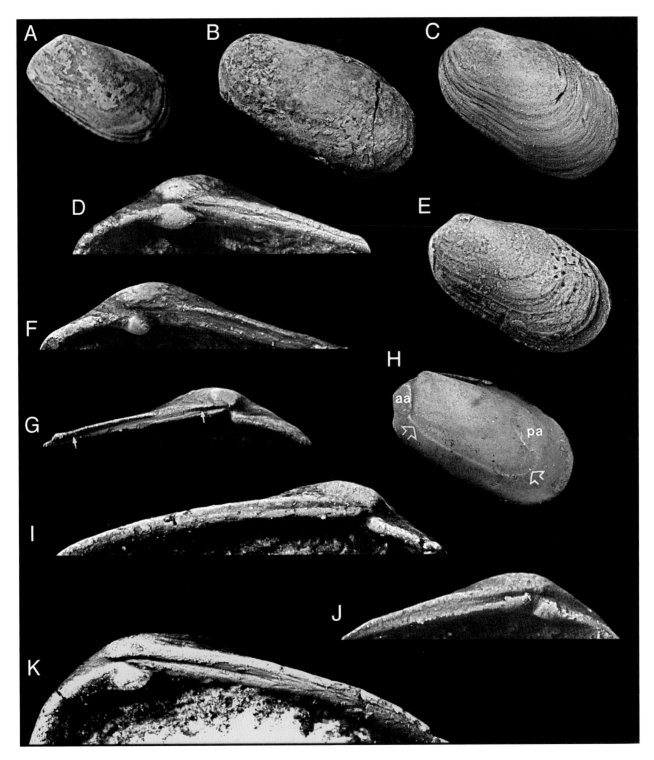

Fig. 50. Mimerodonta njordi n.sp. □A. External lateral view of left valve, RMMo 21707, ×3.0. □B. External lateral view of left valve, RMMo 16363, ×1.9. □C. External lateral view of left valve, holotype, RMMo 16335, ×2.2. □D. Hinge of right valve, RMMo 21504, ×9.1. □E. External lateral view of left valve, RMMo 152516, ×2.5. □F. Hinge of right valve, RMMo 152530, ×9.3. □G. Hinge of left valve, note ligament groove (at arrows), RMMo 152529, ×7.9. □H. Lateral view of internal internal mould of left valve showing muscular impressions of anterior adductor muscle scar (aa), posterior adductor muscle scar (pa), and pallial line (at arrows), LO 6290t, ×2.4. □I. Hinge of left valve, same as E, ×9.3. □J. Hinge of left valve, RMMo 152517, ×8.3. □K. Hinge of right valve, RMMo 152518, ×7.5. All specimens from the Whitcliffian Burgsvik Beds, at Burgsvik, except for H which is from Valar 1, Gotland.

as having hinge teeth subparallel to the dorsum and overlapping one another posteriorly (Pojeta *et al.* 1986).

In the two present species, *Radiatodonta* sp. 1 and *Radiatodonta* sp. 2, the hinge teeth are dorsally convergent.

Radiatodonta sp. 1

Figs. 16D, 27, 51B, D

Material. – A right valve, RMMo 158325. Fig. 51B, D.

Description. – Shell large (maximum length ca. 70 mm) robust, diagonally elliptical, inequilateral, well inflated, margins even, shell surface smooth with concentric growth lines, beaks subanterior, small, close together, located slightly above dorsal margin; shell maximum convexity in posterior half and at about mid-height; no escutcheon; no lunule; dorsal margin long, slightly convex; anterior margin short, lobate; posterior margin long, probably evenly rounded; ventral margin long, sinuous; length of specimen 60 mm (incomplete), height of specimen about 25 mm. Hinge line convex, hinge plate strong with a robust, diagonal cardinal tooth just below beak, flanked by at least two small elongate teeth; posterior part of dorsal margin extends in tooth-like projection; large conspicuous, deep circular anterior adductor muscle scar, deep circular pedal–byssal muscular scar between anterior adductor muscle scar and hinge plate.

Comparisons. – *Radiatodonta* sp. 1 differs from *Radiatodonta* sp. 2 in having a more sinuate ventral margin, being more gibbous, and in lacking a distinct umbonal ridge.

Ecology and habitat. – This species is characterized by a conspicuously sinuated ventral margin suggesting the presence of a byssus. It most probably was a slow burrower and is assumed to have had an endo-byssate life position.

Radiatodonta sp. 1 was found in oolitic limestone indicating an original substrate of coarse oolitic sand.

Occurrence. – Ludlovian (Whitcliffian) Burgsvik Beds, locality unknown, Gotland.

Radiatodonta sp. 2

Figs. 23:15, 25, 27, 51A, C

Material. – A right valve, RMMo 158259. Fig. 51A, C.

Description. – Shell large (maximum length ca. 65 mm), robust, diagonally elliptical, inequilateral, moderately inflated, margins even, shell surface smooth with concentric growth lines; beaks subanterior, small, close together, located slightly above dorsal margin; umbonal ridge prominent in proximal part; maximum convexity at about mid-length and mid-height of shell; no escutcheon; no lunule; dorsal margin long, slightly convex; anterior margin short, narrow; posterior

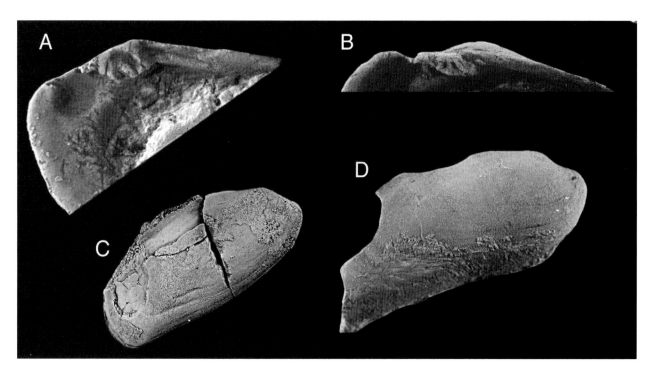

Fig. 51. □A. *Radiatodonta* sp. 2. Hinge of right valve, RMMo 158259. Whitcliffian Burgsvik Beds at Gansviken, Gotland. □B. *Radiatodonta* sp. 1. Hinge of right valve, RMMo 158325, ×3.0. Whitclifian Burgsvik Beds, probably at Burgsvik, Gotland. □C. *Radiatodonta* sp. 2. External lateral view of right valve, same as A, ×0.9. □D. *Radiatodonta* sp. 1. External lateral view of incomplete right valve, same as B, ×1.2.

margin long, evenly rounded; ventral margin long, in anterior part convex, in posterior part sinuated; total length of measured specimen 65.3 mm: height 31.7 mm; width 10.0 mm. Hinge line convex, hinge plate strong with at least two dorsally convergent teeth and two sockets; posterior adductor muscle scar in antero-dorsal position, close to dorsal margin; deep subcircular anterior adductor muscle scar.

Comparisons. – *Radiatodonta* sp. 2 differs from *Radiatodonta* sp. 1 in the following way. The ventral margin of *Radiatodonta* sp. 1 is concave while that of *Radiatodonta* sp. 2 is convex in its anterior part and concave only in its posterior part. *Radiatodonta* sp. 2 also exhibits an umbonal ridge, in contrast to *Radiatodonta* sp. 1, and its shell is more compressed than that of *Radiatodonta* sp. 1. *Radiatodonta* sp. 2 differs from ?*Colpomya lokei* by its conspicuously sinuated posterior part of the ventral margin, sharper umbonal ridge, lower umbo, and smaller beak.

Ecology and habitat. – This species exhibits great similarities in shell morphology to extant endo-byssate mytilids, in that it has an elongate, compressed shell, maximum convexity above mid-height, and the angle between the line through the ligament and the one through the centre of the two adductor muscle scars is almost the same as in *Modiolus modiolus* (see Stanley 1972, p. 172, Text-fig. 4C). *Radiatodonta* sp. 2 has a decidedly compressed shell and is thus considered to have been a rapid burrower (Fig. 23:15). It is suggested that this species had a semi-infaunal, endobyssate life habit (Fig. 13).

Radiatodonta sp. 2 was found in a coarse oolitic limstone, the original sediment being coarse, oolitic sand.

Occurrence. – Ludlovian (Whitcliffian) Burgsvik Beds at Gansviken, Gotland.

Genus *Goniophora* Phillips, 1848
Fig. 36

Type species. – *Goniophora cymbaeformis* Sowerby, 1839.

Emended diagnosis. – Shell equivalve, strongly inequilateral, posteriorly elongated, rather inflated with strong, angular umbonal ridge extending from beak diagonally to inferior angle; beaks small, closely incurved, prosogyrate; umbo prominent; margins even; external opisthodetic ligament, possibly with additional internal portion; comarginal growth lines, in some cases with radial elements on ventral half of shell; hinge plate small, fairly strong, supported by anterior and posterior ridge forming a septum anteriorly and a slender plate or flange posteriorly; hinge either with a single tooth in left valve and a corresponding socket in right, or each valve having one conspicuous cardinal tooth in ventral part of hinge and one marginal tooth (in some cases two) in dorsal part of hinge plate, teeth diagonally arranged; no lateral elements of the hinge; anterior and posterior adductor muscle scars observed; pallial line integripalliate.

Systematic position. – The concept of the genus *Goniophora* Phillips, 1848 (in Phillips & Salter 1848), has long been debated. The original erection of this genus was not accompanied by any diagnosis, and the hinge characters of the type species *Goniophora cymbaeformis* (Sowerby, 1839) are still unknown.

The features in the original illustration of *Goniophora cymbaeformis* (Sowerby 1839, Pl. 3:10; sampled from the Downtonian Long Quarry Beds at Chapel Horeb east of Llandovery, South Wales), which several authors have interpreted as hinge teeth (Fig. 52I herein), are in fact two narrow ridges (cf. identical structures in *Goniophora onyx*; Fig. 52E herein).

The taxonomic affinities of *Goniophora* have been discussed elsewhere (Liljedahl 1984, pp. 63–66), and, to sum it up, there seem to be two different kinds of hinges described in '*Goniophora*-like' species (with the typical umbonal ridge), viz. one equivalved type (equal number of teeth and sockets in each valve; Fig. 52F, H), and one inquivalved type (uneven number of hinge teeth in the two valves).

Based on external features, an array of shells with a diagonal umbonal keel have been assigned to *Goniophora*. Except for the diagonal keel, they may exhibit pronounced dissimilar shape and sculpture. McLearn (1918, p. 140) erected the genus *Cosmogoniophora* (hinge unknown) for *Goniophora*-like shells with radial striae, from the Upper Silurian of North America (type species *Goniophora bellula* Billings). Dahmer (1936, p. 23) referred shells similar to *Goniophora*, also with fine radial costellae and unknown hinge, from the Lower Devonian of Germany, to the new genus *Tylophora* (type species *Goniophora concentrica* Drevermannn). Isberg (1934, p. 202) considered *Goniophora*-like shells without hinge teeth from the Upper Ordovician from Sweden to belong to his new genus *Goniophorina*, and *Cosmogoniophora*-like shells without hinge teeth from the same beds belong to the new subgenus *Goniophorina* (*Cosmogoniophorina*) (Isberg 1934, p. 207). Specimens of Isberg's 'tooth-less' taxa were prepared out of an extremely dense and hard, crystalline limestone. I have studied the original specimens and cannot claim with certainty whether or not these shells had hinge teeth prior to preparation. No specimens of *Goniophorina* (*Cosmogoniophorina*) have any diagnostic radial shell sculpture. Isberg heavily embellished the photographs of his plates.

In the present material two of the species referred to *Goniophora* show radial sculpture, viz. *Goniophora acuta* and *Goniophora brimeri*. Unfortunately, no hinge structure has been observed in either of these, and, consequently, their generic affinity cannot be confirmed.

'*Goniophora*-like' specimens with unknown hinge characters may belong to two or more different genera. External sculpture and shell form, on the other hand, are not considered as of generic but of specific significance. Thus, *Cosmogoniophora* McLearn, *Tylophora* Dahmer, *Goniophorina* Isberg, and *Goniophorina* (*Cosmogoniophorina*) Isberg are suggested to be synonyms of *Goniophora* Phillips.

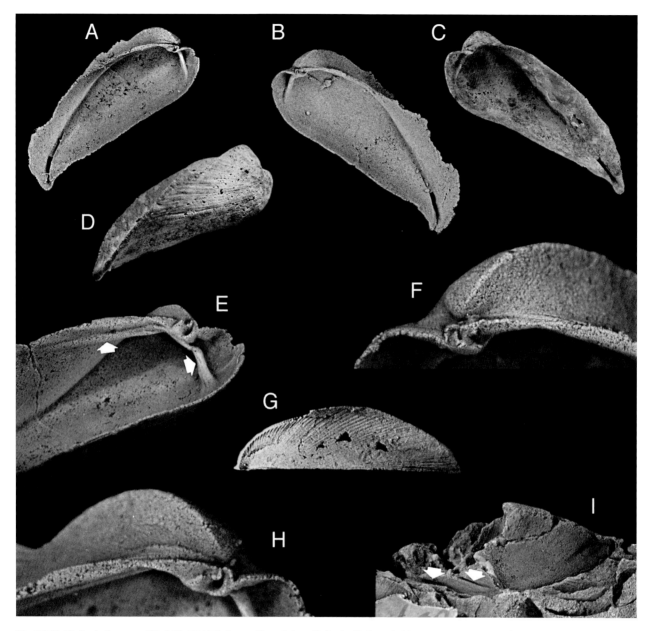

Fig. 52. □A–H. *Goniophora onyx* Liljedahl, 1984. □A. Internal lateral view of left valve (holotype), SGU Type 3736, ×1.9, sample G79-90LJ. □B. Internal lateral view of right valve (holotype), SGU Type 3735, ×1.9, sample G79-90LJ. □C. Internal lateral view of right valve, SGU Type 8372, ×2.0. □D. External lateral view of right valve, ×2.0, same as C. □E. Internal ventro-lateral view of left valve (holotype) showing anterior and posterior reinforcement ridge, respectively (at arrows), ×3.2. □F. Hinge of right valve (holotype), ×6.7. □G. External dorsal view of right valve, same as C, D, ×2.1. □H. Hinge of left valve (holotype), ×7.7. All specimens from the Homerian Halla Beds at Möllbos 1, Gotland. □I. *Goniophora cymbaeformis* (Sowerby, 1839). External dorsal view of incomplete right valve (holotype), illustrated by Sowerby (1839, Pl.3:10a), British Geological Survey Geological Society collection no. 6689, showing two posterior reinforcement ridges, seen as slits in the mould (at arrows, see text), ×1.0. Chapel Horeb, Pridoli, Long Quarry Beds.

Goniophora onyx Liljedahl, 1984

Figs. 16F, 24, 27, 52A–H

Synonomy. – □1984 *Goniophora onyx* n.sp. – Liljedahl, 63–72, Figs. 29–31.

Holotype. – One right valve, SGU Type 3735 (Fig. 52B, F) and one left SGU Type 3736 (Fig. 52E, H), both from sample G79-90LJ.

Type stratum. – Halla Beds, Late Wenlockian.

Type locality. – Möllbos 1, Gotland.

Material. – Forty-two silicified single valves, eight of which with fairly well preserved hinge, and one internal mould with parts of shell material preserved as calcium carbonate. RMMo 21745.

Diagnosis (from Liljedahl 1984). – *Goniophora* with dorsal margin slightly convex, posterior margin evenly rounded, anterior margin lobate, ventral margin straight except for its proximal end, where it has a claw-like appearance laterally; hinge plate with one large cardinal tooth and one smaller marginal tooth with corresponding sockets separating them, obliquely placed in relation to hinge line; two ridges, or septa, supporting and originating from underneath the hinge plate, one anteriorly and the other posteriorly.

Description. – See Liljedahl (1984). 2W/L = 0.56, H/2W = 0.71, H/L = 0.41 (N = 2).

Comparisons. – Among species with well-documented internal features the Devonian species *Goniophora secans* Barrande, 1881, seems to be most closely related to *Goniophora onyx* (see Liljedahl 1984). However, it differs externally in the following way. *Goniophora secans* exhibits a more developed posterior end at the point of maximum extension of the shell (in *G. onyx*, this is in the anterior part), a less developed anterior lobe, and an umbonal ridge which is erect except for its proximalmost part, where it is dorsally inclined. The umbonal ridge of *G. onyx* is dorsally inclined from mid-length of the shell. Furthermore, the claw-like posterior end present in *G. onyx* is lacking in *G. secans*.

  Goniophora onyx differs from *G. bragei* and *G. acuta* by the claw-like shape of the shell, and from the latter also by its lobate anterior end, and in having a C-shaped umbonal ridge in the right valve, as seen in lateral view. The umbonal ridge of *G. acuta* is comma-shaped in the right valve.

  In its claw-like lateral silhuette, *Goniophora onyx* shows similarity to the Devonian *G. schwerdi* Beushausen (1895, p. 206, Pl. 17:22–30) which, however, has a much shorter shell.

Ecology and habitat. – *Goniophora onyx* has a thin, comparatively broad unornamented shell with a distinct umbonal keel. When seen in lateral view, the keel divides the shell in a large ventral part and a smaller dorsal part consisting of the posterior margin only. The life habit of this species is suggested to have been semi-infaunal with the keel at the sediment/water interface, and, thus, with most of the shell buried (Fig. 24). Specimens of *Goniophora onyx* were isolated from a fine-grained argillaceous limestone which was once a soft mud. In such an environment the broad shell of this species would have produced a good physical stability ('snow shoe' effect; Thayer 1974). If a byssus was present, there was no need for it to be very strong. *G. onyx* has only been found as disarticulated valves. The disarticulation is probably due to intense bioturbation (Liljedahl 1985).

Occurrence. – Wenlockian Halla Beds at Möllbos 1, Ludlovian Hemse Beds at Hammarudd, Gotland.

Goniophora bragei n.sp.

Figs. 24, 27, 53B–D, H

Synonymy. – □?1895 *Goniophora schwerdi* n.sp. – Beushausen, p. 206, Pl. 17:22–30.

Derivation of name. – After *Brage*, poet of Valhall and son of the god Woden, in the old Scandinavian mythology.

Holotype. – A complete left valve, RMMo 158744. Fig. 53D.

Type stratum. – Probably Burgsvik Beds.

Type locality. – Unknown, probably Burgsvik, Gotland.

Material. – Nine specimens, internal moulds with more or less complete shells preserved as calcium carbonate.

Diagnosis. – Shell small, dorsal margin convex, posterior margin straight, anterior margin pointed and lobate, ventral margin almost straight; hinge plate with a single(?) cardinal tooth and a socket in the right valve and corresponding features in the left; hinge plate reinforced by ridge anteriorly.

Description. – Shell small (maximum length ca. 17 mm), subtriangular (Fig. 53D), inflated, margins even, beaks small, close together, strongly incurved, prosogyrate, high erect umbonal ridge, angular, becoming conspicuously narrow and high-crested posteriorly (Fig. 53D); maximum convexity at about mid-height and in upper half of the shell (Fig. 53H); shell surface smooth with comarginal growth lines; dorsal margin convex; anterior margin short, rounded, lobate; posterior margin almost straight, truncated; ventral margin long, almost straight; hinge line convex, hinge plate with at least one

---

Fig. 53. □A. *Goniophora brimeri* n.sp. Dorsolateral view of internal cast of articulated specimen showing hinge reinforcement ridge (r) as a slit, hypertrophied posterior adductor muscle (pa) scar, and posterior byssal muscle scar (pbr), RMMo 21722, ×2.6. Ludlovian Hemse Beds at Östergarn. □B. *Goniophora bragei* n.sp. External lateral view of right valve, RMMo 158722, ×1.9. Whitcliffian Burgsvik Beds at Burgsvik. □C–D. *Goniophora bragei.* □C. External lateral view of left valve, RMMo 21656, ×2.3. Whitcliffian Burgsvik Beds at Burgsvik. □D. External lateral view of left valve (holotype), RMMo 158749, ×2.2. Whitcliffian Burgsvik Beds at Rovalds. □E–G. *Goniophora brimeri* n.sp. □E. Lateral view of left valve of internal cast of articulated specimen, same as A, ×2.1. □F. Dorsal view of same specimen as A, E, ×1.9. □G. External lateral view of left valve of internal cast of articulated specimen, RMMo 21741, ×1.7, Wenlockian Visby Beds at Visby. □H. *Goniophora bragei.* External dorsal view of right valve, RMMo 158722, ×2.4. Whitcliffian Burgsvik Beds at Burgsvik. □I. *Goniophora alei* n.sp. Internal lateral view of left valve (holotype), RMMo 158743, ×4.4. Unknown locality of the Burgsvik Beds. □J. *Goniophora tyri* n.sp. External lateral view of right valve (holotype), RMMo 21682, ×1.7. Ludlovian Hemse Beds at Linde Klint. □K. *Goniophora alei.* External lateral view of left valve (holotype), same as I, ×3.0. □L. *Goniophora alei.* External dorsal view of left valve, (holotype), same as I and K, ×3.9. □M. '*Goniophora*' sp. 1. Internal lateral view of left valve, RMMo 158746, ×1.65. Locality unknown. □N. '*Goniophora*' sp. 1. External lateral view of left valve (same as M), ×1.2. □O. *Goniophora tyri.* External dorsal view of incomplete left valve showing conspicuous 'finger print' shell sculpture, RMMo 21753, ×2.5. Ludlovian Hemse Beds at Åsa träskbacke. All localities in Gotland.

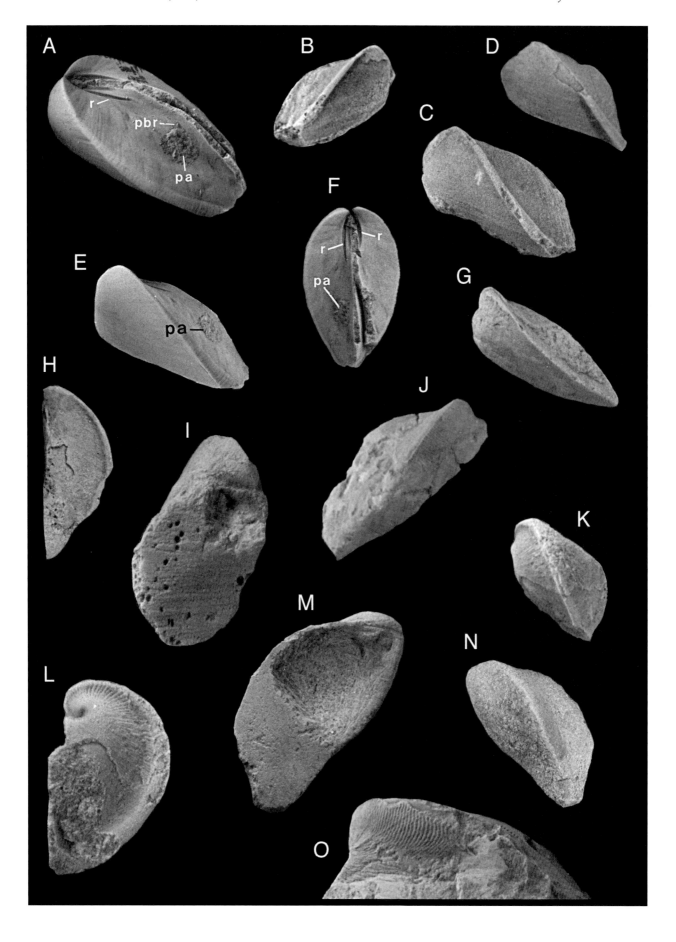

cardinal tooth and socket in each valve; hinge reinforced by posterior ridge; anterior septum not observed; deeply impressed anterior adductor muscle scar present, no other internal features observed, $2W/L = 0.69$, $H/2W = 0.63$, ($n = 5$), $H/L = 0.47$.

Comparisons. – *Goniophora bragei* seems to be most closely related to the Devonian *G. schwerdi* Beushausen (1895, p. 206, Pl. 17:22–30; see also Maillieux 1937, p. 134, Pl. 10:7), from which it differs by its almost straight ventral margin. *G. schwerdi* has a pronounced sinuated ventral margin resulting in a hook-like appearance in lateral view.

Goniophora bragei also shows some similarity to the Bohemian (Upper Silurian–Lower Devonian) *G. testis* Barrande (1881, Pl. 261:V) but differs from this by its straight ventral margin compared with the conspicuously convex ventral margin of *G. testis*.

Goniophora bragei, with length/height ratio of 1.92, also resembles the Bohemian (Silurian–Devonian) *G. media* Barrande (1881, Pl. 357:9), which, however, is decidedly longer (L/H = 2.73).

Goniophora bragei shows some similarity to the Silurian *G. perangulata* Hall, 1870 (in Hall 1885, p. 293, Pl. 34:2), but the umbonal ridge of that species is more dorsally inclined in the anterior half of the shell and, furthermore, it has a distinct mesial sulcus.

The Ordovician *Goniophorina solvens* Isberg, 1934, is externally reminiscent of *Goniophora bragei*, but its internal features are unknown and thus, its systematic relationship is difficult to establish.

Ecology and habitat. – *Goniophora bragei* has a small and unsculptured shell characterized by its sharp-edged umbonal keel. It has been found only as single valves in oolitic limestones and sandstones, indicating an unstable habitat. *G. bragei* is believed to have had a semi-infaunal life position (Fig. 24) and been provided with a byssus.

Occurrence. – Ludlovian (Whitcliffian) Burgsvik Beds at Burgsvik, Gotland.

Goniophora brimeri n.sp.

Fig. 27, 53A, E–G

Derivation of name. – After *Brimer*, a by-name of Mimer, the infernal ruler and guardian of the spring of wisdom, from the old Scandinavian mythology.

Holotype. – A complete articulated steinkern, RMMo 21722. Fig. 53A, E–F.

Type stratum. – Ludlovian Hemse Beds.

Type locality. – Östergarn, Gotland.

Material. – Fifteen specimens preserved as internal moulds or with shell material.

Diagnosis. – *Goniophora* with posterior margin evenly rounded, ventral margin straight, non-lobate and truncate anterior end, S-shaped umbonal ridge; hinge with one anterior septum and one posterior ridge.

Description. – Shell medium-sized (maximum length ca. 28 mm), lanceolate (Fig. 53E, G), inflated, margins even; shell surface smooth with comarginal growth lines, ventral half also with radial riblets; beaks small, close together, incurved, prosogyrate; umbonal ridge S-shaped in lateral view, angular, lacking crest, dorsally inclined; maximum convexity at mid-length and at mid-height of shell; dorsal margin short, strongly convex; anterior margin short, truncate, non-lobate; posterior margin long, evenly rounded; ventral margin straight to slightly concave; posterior reinforcement ridges observed (Fig. 53A, E–F); posterior adductor muscle scar with associated byssal–pedal retractor muscle scar observed (Fig. 53A, E), $2W/L = 0.61$, $H/2W = 0.65$, ($n = 5$), $H/L = 0.48$ ($n = 1$).

Comparisons. – *Goniophora brimeri* is most easily distinguished from *G. bragei*, with which it seems to be most closely related, by its non-lobate anterior end, its S-shaped umbonal ridge, and by the fact that it lacks a crest on the umbonal ridge.

Ecology and habitat. – *Goniophora brimeri* has a comparatively broad shell with reduced anterior lobes and a conspicuous pedal–byssal muscle scar, indicating the presence of a byssus. This species has been found in fine-grained limestones and siltstones, and it probably had a semi-infaunal life position. Most specimens of *G. brimeri* are preserved as articulated specimens, which strengthens the assumption that the fine-grained sediments, in which they have been found, represent their original habitats.

Occurrence. – Llandoverian Lower Visby Beds or Wenlockian Upper Visby Beds at Visby; Wenlockian Slite Beds at Vall kanal; Ludlovian Hemse Beds at Petesvik, Gotland; Klinta Formation at Klinta and Ramsåsa, Scania.

Goniophora tyri n.sp.

Fig. 24, 27, 53J, O

Derivation of name. – From the old Scandinavian Aesir war god *Tyr*, son of Woden.

Holotype. – An almost complete right valve, RMMo 21682. Fig. 53J.

Type stratum. – Ludlovian Hemse Beds.

Type locality. – Linde Klint, Gotland.

Material. – Nine disarticulated incomplete specimens preserved with shell material as calcium carbonate.

Diagnosis. – *Goniophora* with evenly posterior margin rounded, slightly concave ventral margin, S-shaped umbonal ridge, conspicuous transverse 'finger print' sculpture, and hinge with one cardinal tooth and socket in each valve.

Description. – Shell medium-sized (maximum length ca. 35 mm), sublanceolate (Fig. 53J), inflated, anterior part lobate, shell surface with conspicuous transversly undulating costellae (Fig. 53O); beaks small, close together, prosogyrate; dorsally inclined, S-shaped (from a lateral view) umbonal ridge; dorsal margin long, convex; anterior margin short, lobate; posterior margin long, evenly rounded; ventral margin slightly concave; hinge line convex, hinge with one diagonal cardinal tooth and socket in each valve; no other internal features observed. 2W/L = 0.56 ($n = 3$).

Comparisons. – *Goniophora tyri* differs from all other described *Goniophora* species by its conspicuous rugose transverse shell sculpture. It shows close external similarity to *Goniophora* sp. from the late Ordovician (Ashgill) of Ireland, illustrated by Tunnicliff (1982, p. 80, Pl. 12:15). The hinge of that unnamed species of *Goniophora*, is however, unknown.

Ecology and habitat. – This species is primarily distinguished by its unique shell sculpture of transverse finger-print-like concentric costellae. Possibly, these strengthened the shell, but they may also have assisted in keeping the valves in a stable life position.

The umbonal keel divides the shell in two subequal halves, resulting in a life position with more than half the shell above the sediment surface when positioned with the umbonal keel parallel to the surface. The species has been found in siltstones and coarse-grained limestones, and its fairly broad shell suggests a semi-infaunal mode of life, as presented in Fig. 24. No muscular impressions have been observed, but probably it had a byssus to maintain a stable life position, in view of the fact that such a great part of the shell is likely to have been above the sediment surface.

Occurrence. – Ludlovian Hemse Beds at Linde klint, Gotland.

Goniophora alei n.sp.

Fig. 27, 53I, K, L

Derivation of name. – After the old Nordic name *Ale*.

Holotype. – An incomplete left valve, RMMo 158743 with shell preserved as calcium carbonate. Fig. 53I, K–L.

Type stratum. – Unknown.

Type locality. – Unknown, Gotland.

Material. – Only the holotype.

Diagnosis. – *Goniophora* with a conspicuously gibbous valve, a sharply defined umbonal ridge and an involute beak.

Description. – Shell small (length 13 mm), lanceolate (Fig. 53K), gibbous, margins even, shell surface with comarginal growth lines, beaks conspicuously small, close together,

prosogyrate, involute (Fig. 53L); umbo high, umbonal ridge S-shaped in the left valve, with erect crest; maximum convexity at about mid-height and mid-length of shell; dorsal margin convex; anterior margin almost non-lobate; posterior margin convex; ventral margin slightly convex: no internal features observed. 2W/L = 0.84, H/L = 0.47 ($n = 1$).

Remark. – No other species of the genus exhibits such a strongly incurved beak, which in this case is best described as involute, with the possible exception of *G. acuta* Sandberger (in Beushausen 1895, p. 211, Pl. 17:1–3), which otherwise is quite dissimilar to the present species.

Comparisons. – This species is easily distinguished from most known species of *Goniophora* by its strongly involute beak.

Ecology and habitat. – This species exhibits the broadest shell of all the species of *Goniophora* dealt with here. It shows similarity to living endo-byssate mytilids in exhibiting a triangular cross section with maximum width at mid-height of shell. Thus, it may have had a semi-infaunal life habit and have been attached with a strong byssus.

Goniophora alei has been found in oolitic limestones suggesting an original high-energy habitat. In such an environment, an effective byssus is a prerequisite to maintain a stable life position for a small, relatively thick-shelled endobyssate bivalve.

Occurrence. – Locality and horizon unknown, Gotland.

Goniophora acuta Lindström, 1880

Figs. 24, 27, 54A–H

Synonymy. – □*non* 1850 *Cypricardia? acuta* n.sp. – Sandberger, p. 263, Pl. 27:12. □*non* 1870 *Sanguinolites acutus* Hall – Hall, p. 37. □*non* 1877 *Sanguinolites acutus* Hall – Miller, p. 202. □*non* 1877 *Goniophora acuta* Hall – Miller, p. 192. □v*1880 *Goniophora acuta* Lindström – *in* Angelin & Lindström, p. 19, Pl. 19:23–26. □*non* 1881 *Goniophora acuta* Barrande – Barrande, p. 81. □*non* 1885 *Goniophora acuta* Hall – Hall, p. 295, Pl. 43:1–3.

Holotype. – A complete, deformed articulated specimen, RMMo 150417, preserved as calcium carbonate. Fig. 54A, D.

Type stratum. – Klinta Formation, Late Ludlovian.

Type locality. – Klinta, Scania.

Material. – Four articulated specimens showing external features only, preserved as calcium carbonate.

Emended diagnosis. – *Goniophora* with long dorsoposterior margin and subparallel anteroventral margin, conspicuously narrow and elongated and with high inflation, erect umbonal ridge.

Description. – Shell large (maximum length ca. 50 mm), lanceolate (Fig. 54B), conspicuously elongate, margins even,

shell surface with conspicuous comarginal growth threads (Fig. 54B, C, G), ventral half also with radial riblets (Fig. 54F); beaks small, terminal, close together, slightly prosogyrate; umbo high with angular, erect umbonal ridge slightly curved; maximum convexity in anterior and lower part of shell; dorsal margin long, almost straight; anterior margin short, almost straight; posterior margin long, evenly rounded; ventral margin slightly convex; no internal features observed. $2W/L = 0.45$, $H/2W = 0.65$ ($n = 4$), $H/L = 0.22$ ($n = 1$)

Comparisons. – This species is easily distinguished from other species of the genus by its narrow, elongated lateral shell shape, which is particularly emphasized by the acute angle formed by the dorsal and anterior margin, respectively.

Ecology and habitat. – The shell of *Goniophora acuta* is conspicuously flattened and broad. The straight diagonal keel divides the shell into two equally large parts. In addition to the fine concentric shell pattern, the ventral half also exhibits a coarser radial sculpture, possibly strengthening the shell and also stabilizing the animal in its reclining life position. The beak is small and pointed and unsuitable for attachment of strong byssal muscles. *G. acuta* has been found in fine-grained, argillaceous limestones and siltstones. One specimen has been collected in situ with the diagonal keel parallel to the bedding plane (Fig. 54H). Specimens have only been found articulated, indicating an original low-energy habitat. In substrates of such habitats, there was no need for the flat, low-profile shell of *G. acuta* to have a byssus in order to maintain a stable life position on the sediment surface (Fig. 24).

Occurrence. – Wenlockian Slite Beds at Valbytte 3; Ludlovian Hemse Beds at Petesvik, Gotland, and Klinta Formation at Klinta, Scania.

Fig. 54. □A–H *Goniophora acuta* Lindström, 1880. □A. External dorsal view of articulated specimen (holotype; same as D), anterior facing left, RMMo 150417, ×1.5. Whitcliffian Klinta Formation at Klinta, Scania. □B. External lateral view of right valve of articulated specimen. Note radial sculpture in ventral half of valve, RMMo 158729, ×1.3. Ludlovian Hemse Beds at Petesvik, Gotland. □C. External dorsal view of articulated specimen, same as B, ×1.3. □D. External lateral view of left valve of holotype (specimen somewhat dorsoventrally compressed, owing to compaction), ×1.3. □E. External anterior view of articulated specimen (same as A, D), ×1.3. □F. External ventral view of articulated specimen, note radial shell sculpture (same as C, E). □G. Shell surface sculpture of dorsal part of shell. □H. External lateral view of articulated specimen in life position (at arrows) with umbonal ridge parallel to bedding plane (substrate bed eroded), LO 6346t, ×0.9. Whitcliffian Klinta Formation at Klinta, Scania. □I. 'Goniophora' gymeri n.sp., external lateral view of right valve, RMMo 158748, ×1.2. Locality on Gotland unknown. □J. 'Goniophora' gymeri. External dorsal view of right valve, same as I, ×1.2. □K. *Mytilarca* sp. Internal lateral view of left valve (see discussion in text), RMMo 8146, ×1.3. Klinteberg Beds at Öjlemyrs kanal, Gotland. □L. 'Goniophora' gymeri. Internal lateral view of right valve, same as I and J, ×1.3. □M. 'Goniophora' gymeri. External lateral view of right valve (holotype), RMMo 158726, ×1.3. Wenlockian Slite Beds at Bjärs, Gotland.

'Goniophora' gymeri n.sp.
Figs. 24, 27, 54I–J, L–M

Systematic position. – The robust hinge plate of this species is distinct from the small hinge plate characteristic of typical *Goniophora* such as e.g., *G. onyx*. Most probably 'G.' gymeri belongs to another, so far unrecognized, genus, and is only tentatively placed in *Goniophora*.

Derivation of name. – After *Gymer*, a giant in the old Scandinavian Mythology.

Holotype. – An incomplete left valve, RMMo 158726. Fig. 54M.

Type stratum. – Wenlockian Slite Beds, Gotland.

Type locality. – Bjärs, Hejnum, Gotland.

Material. – Four fragmented specimens preserved with calcium carbonate shells.

Diagnosis. – Dorsal margin long, convex; anterior margin short, pointed; anterior and middle part of ventral margin straight; beak small, terminal; hinge plate triangular, robust, anterior adductor muscle scar on umbonal shelf.

Description. – Shell large (maximum length ca. 45 mm), lanceolate elongate (Fig. 54I, M), inflated; margins even; beaks extremely small, close together, incurved, prosogyrate; C-shaped umbonal ridge in a right valve; maximum convexity above mid-height and at about mid-length of shell; shell surface smooth with comarginal growth lines; dorsal margin long, convex; anterior margin short, pointed, forming acute angle with the dorsal margin; posterior margin not observed; ventral margin long, anterior and middle part almost straight, posterior part not observed; hinge line convex, hinge plate robust, triangular; hinge teeth eroded; anterior adductor muscle scar on umbonal shelf (Fig. 54L): no further internal features observed. $2W/L = 0.79$, $H/2W = 0.55$ ($n = 4$), $H/L = 0.75$ ($n = 1$).

Comparisons. – 'Goniophora' gymeri is distinguished from other *Goniophora* species by its large size, robust hinge, and its terminal beak. The terminal beak closely resembles that of, e.g., *Mytilarca*. 'G.' gymeri seems to be most closely related to 'G.' sp. 1 but differs from this by its C-shaped and more dorsally inclined umbonal ridge in the proximal part, its more acute angle formed by the anterior and dorsal margins, and its higher umbo.

Ecology and habitat. – The shell of 'Goniophora' gymeri is large. Its umbonal keel is strongly sigmoidal and the beak terminal. The hinge plate is extremely thick and the anterior adductor muscle scar is housed on an umbonal shelf (in contrast to the remaining species of this genus except for 'G.' sp. 1 herein). It shows similarities in external morphology to Recent epifaunal mytiliids as well as to the Pennsylvanian *Promytilus* Newell, 1942, which is believed to have been epifaunal, and which, in fact, is considered to be the ancestor of *Mytilus* (Newell 1942, p. 37).

Just like *Promytilus*, 'Goniophora' *gymeri* and 'G.' sp. 1 are found in coarse sediments which supports the suggestion that the latter two also lived epifaunally and were attached to larger particles (Fig. 24).

Occurrence. – Wenlockian Slite Beds at Bjärs and Storugns, Gotland.

'Goniophora' sp. 1
Figs. 24, 53M–N

Systematic position. – The robust hinge plate of this species is distinct from the small hinge plate characteristic of 'typical' *Goniophora* such as *G. onyx*. 'G.' sp. 1 is closely related to 'G.' *gymeri* and is suggested to belong to the same, as yet unnamed, genus as that species. It is thus only tentatively placed in *Goniophora*.

Material. – One left valve, RMMo 158746, preserved with calcium carbonate shells, Fig. 53M–N.

Description. – Shell large (length 41 mm), thick, robust, lanceolate, margins even, shell surface smooth with comarginal growth lines; beaks small, close together, prosogyrate; umbo low with S-shaped umbonal ridge in a left valve, maximum convexity above mid-height and in posterior part of shell; dorsal margin long, convex; anterior margin short, pointed, forming an acute angle with dorsal margin; posterior margin long, truncated; ventral margin S-shaped; hinge line almost straight, hinge plate triangular, robust, hinge teeth eroded; anterior adductor muscle scar on umbonal shelf: no further internal features observed, 2W/L = 0.65, H/2W = 0.81, H/L = 0.52 ($n = 1$).

Comparisons. – This species seems to be most closely related to 'Goniophora' *gymeri*, from which it is distinguished by its lower umbo (Fig. 53M), less dorsally inclined umbonal ridge in the proximal part, lesser acute angle formed by the anterior and dorsal margins, and its sinuated ventral margin and S-shaped umbonal ridge.

Ecology and habitat. – The only specimen of 'Goniophora' sp. 1 was found in a coarse-grained limestone suggesting a high-energy habitat. It is believed to have had a strong byssus, and probably it had an epifaunal life habit (Fig. 24).

Occurrence. – Unknown, Gotland.

Subfamily Modiolopsinae n.subf.
Fig. 36

Diagnosis. – Modiomorphidae without hinge teeth.

Type genus. – *Modiolopsis* Hall, 1847.

Genera. – *Corallidomus* Whitfield, 1893; *Modiolopsis* Hall, 1847; *Orthodesma* Hall & Whitfield, 1875; *Pholadomorpha* Foerste, 1914; *Whiteavesia* Ulrich, 1893; and others.

Genus *Modiolopsis* Hall, 1847
Fig. 36

Type species. – *Modiolopsis modiolaris* (Conrad, 1841)

Diagnosis. – Equivalved, inequilateral, modioliform shell which expands posteriorly; opisthodetic, parivincular ligament, hinge plate narrow, edentulous; surface smooth; subisomyarian musculature, deeply impressed anterior muscle scar, larger but fainter posterior adductor muscle scar; several accessory pedal and byssal muscle scars; pallial line non-sinuate, continuous.

Modiolopsis alvae n.sp.
Figs. 16E, 23:16, 25, 27, 55, 56

Derivation of the name. – After the old Nordic name *Alva*, feminine of Alvar, meaning earnest.

Holotype. – A complete left valve, LO 6293T, preserved as a silicified replica, sample G83-3LL. Fig. 55A.

Type locality. – Gothemshammar 7, Gotland, Sweden.

Type stratum. – Halla Beds c, Homerian, Late Wenlockian.

Material. – One complete left valve specimen (holotype) and a number of fragments of silicified specimens and a few non-silicified calcium carbonate specimens.

Diagnosis. – Shell large, compressed, length twice the height, hinge plate edentulous; myophoric buttress continuing as a narrow ridge posteriorly; anterior adductor muscle scar with uneven posterior limitation, pallial line non-sinuate, formed by punctiform scars in anterior part and radial scars in medium part; conspicuous depression in anterior lobe.

External features. – Shell large (maximum length ca. 36 mm), subelliptical (Fig. 55A–B, E), equivalve, inequilateral, compressed, margins even; total length about twice the height; shell surface smooth with faint concentric increment stops; beaks subanterior, small, close together, prosogyrate, almost in line with dorsal margin; no umbonal ridge, maximum convexity at about mid-length and at mid-height of shell; no escutcheon; no lunule; dorsal margin long, slightly convex; anterior margin straight, almost pointed, forming a conspicuous anterior lobe; ventral margin long, almost straight; posterior margin evenly rounded; total length of measured shells 18.2–36.2 mm ($a = 27.98$ mm; $n = 8$): L/H 1.75–2.12 ($a = 1.93$; $n = 8$); H/2W = 1.40–2.28 ($a = 1.87$. $n = 3$).

Internal features. – Hinge line slightly convex, hinge plate comparatively strong in anterior and central part, edentulous

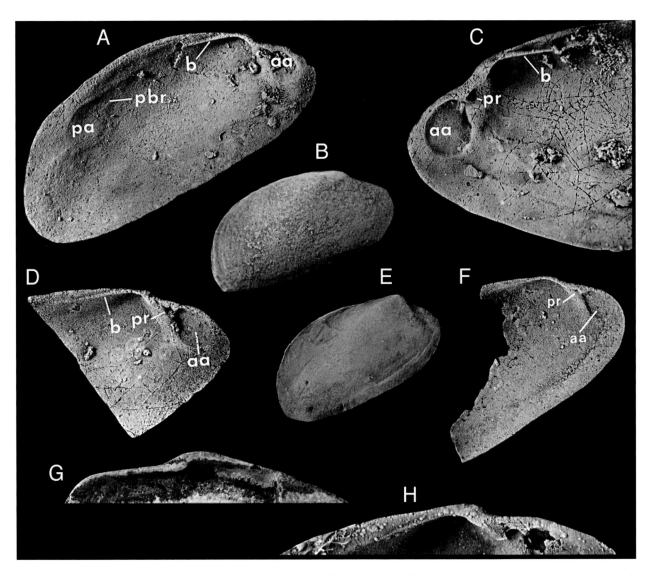

Fig. 55. Modiolopsis alvae n.sp. □A. Internal lateral view of left valve (holotype) showing posterior adductor muscle scar (pa), posterior byssal retractor muscle scar (pbr), buttress (b), and anterior adductor muscle scar (aa), LO 6293T, ×2.4, sample G83-3LL. □B. External lateral view of right valve, RMMo 158260, ×1.7. □C. Internal lateral view of anterior part of right valve showing hinge with buttress (b) and muscular impressions of anterior adductor (aa) and anterior pedal retractor (pr), LO 6294t, ×2,5, sample G83-1LL. □D. Internal lateral view of anterior part of left valve showing hinge with buttress (b), impressions of anterior adductor (aa), and anterior pedal retractor (pr), LO 6297t, ×2.8. □E. Lateral view of internal cast of left valve of articulated specimen, RMMo 18623, ×1.5. □F. Internal lateral view of anterior part of left valve showing anterior pedal retractor muscle scar (pr) and anterior adductor muscle scar (aa), LO 6296t, ×2.9. □G. Hinge of right valve, RMMo 158260, ×3.8. □H. Hinge of left valve, LO 6295t, ×4.2, sample G83-1LL. All specimens from the Wenlockian Halla Beds at Gothemshammar 7, Gotland.

(Fig. 55A, C, G–H); hinge plate continuing posterior to beak in a conspicuous ridge, or buttress, diverging from dorsal margin, and becoming fainter distally (Fig. 55C); large, circular, weakly impressed posterior adductor muscle scar (Fig. 55A) in a posterodorsal position at some distance from the shell margin (No. 1 in Fig. 56); subcircular, large anterior adductor muscle scar (Fig. 55C, D–F), deeply incised, especially in its posterior part (No. 2 in Fig. 56); one deep, medium-sized, circular scar (Fig. 55A, C–D, H) in an anterior position in the umbonal cavity (No. 3 in Fig. 56); one small scar in contact with the anterior adductor muscle scar in a posterodorsal position (No. 4 in Fig. 56); two small scars (Fig. 55C) in contact with the anterior adductor muscle scar in a posterior position (Nos. 5–6 in Fig. 56); hypertrophied anteroposterior part of the posterior adductor muscle scar (No. 7 in Fig. 56); pallial line non-sinuate, running from ventral end of anterior adductor muscle scar to posterior adductor muscle scar with punctiform scars in the anterior part, and radial scars in the medium part; conspicuous depression posterior to anterior adductor muscle scar.

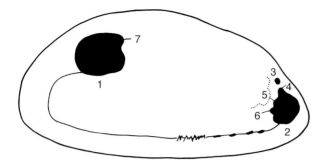

Fig. 56. Modiolopsis alvae n.sp . Maximum number of muscular imprints in order of size. 1: Posterior adductor muscle scar. Note hypertrophied dorsoanterior part indicating byssal retractor muscle scar (no. 7). 2: Anterior adductor muscle scar. 3: Anterior byssal–pedal retractor muscle scar. 4–6: Anterior pedal accessory muscle scars.

Remarks. – The collection of silicified specimens available is small and fragmentary, owing to incomplete silicification and to the fact that the material was most probably fragmented and worn prior to burial. The shells also have been heavily attacked by boring and other destructive organism activity.

Since no articulated specimens have been isolated, the relative convexity of the opposing valves cannot be precisely determined but is suggested to have been approximately equal.

The ridge (buttress) originating from the hinge plate possibly served as a reinforcement of the hinge of a comparatively thin shell (cf. similar structures in the almost contemporaneous *Goniophora onyx* Liljedahl 1984, Fig. 52E herein).

Modiolopsis alvae shows no indication of the presence of any external ligament (nymph or such), nor of any internal one. The dorsal margins of the valves seem to meet and close tightly, suggesting the presence of an internal linear, or 'mytilid', ligament, although no ligamental ridges have been observed. Pojeta (1971, p. 21) discussed the ligament in some well preserved silicified Ordovician modiomorphids. He, too, observed no ligament ridges such as those characteristic of Recent Mytilidae (Pojeta 1971, p. 21, Pls. 13:6–8, 17:1, 3, 11, 13; see also Trueman 1950; Soot-Ryen 1955).

The pattern of muscular impressions, as well as the shell shape of *Modiolopsis alvae*, indicate a byssally attached life habit (see discussion below in the section 'Ecology and habitat'). Although being heavily impressed, the anterior adductor muscle scar is reduced in size compared to the posterior one. The irregular limitation of the anterior adductor muscle scar indicates the presence of two additional scars (Nos. 5–6 in Fig. 56) which, together with the conspicuously deep scar just below the anterior part of the hinge plate (No. 3 in Fig. 56), are believed to be the sites of pedal and byssal–pedal retractor muscles. Also the hypertrophied anterodorsal end of the posterior adductor muscle scar indicates the incisement of a separate muscle, most probably a posterior byssal–pedal retractor muscle (No. 7 in Fig.56).

Systematic position. – *Modiolopsis alvae* exhibits characteristics typical of modioliform modiomorphids, such as obliquely subovate shell shape, anterior reduction of shell, inconspicuous beaks, compressed valves, and complex pattern of accessory muscle scars in the anterior umbonal region. Although the hinge is strenghtened by a buttress, which has not been observed in any other modiomorphid, and the hinge plate is considerable thinner than in *Modiolopsis*, it is tentatively placed in *Modiolosis*.

Also *Whiteavesia* Ulrich exhibits similarities to *Modiolopsis alvae*, such as general shell form (except for its concave ventral margin), and the broadening of the hinge plate below the beak. Furthermore, *Whiteavesia* has large, prominent umbones. Both *Modiolopsis* and *Whiteavesia* have a muscular pattern similar to that of *Modiolopsis alvae*.

Endodesma Ulrich bears some resemblance to *Modiolopsis alvae* in gross shell shape but differs from it in exhibiting a prominent umbo with deep umbonal sulcus, resuting in a sinuated ventral margin, and also in the fact that it lacks a myophoric buttress.

Species of *Orthodesma* Hall & Whitfield and *Psiloconcha* Ulrich also resemble *Modiolopsis alvae*, but these have gaping shells contrary to the latter's tightly fitting shells margins.

Other modiomorphid genera, such as *Modiolodon* Ulrich and *Modiomorpha* Hall & Whitfield, also contain species bearing similarities to *Modiolopsis alvae*, but these possess a strong hinge with conspicuous cardinal tooth or teeth.

Comparisons. – The edentulous *Ectenocardiomorpha acuta* Isberg, 1934, from the Upper Ordovician of Dalarna, Sweden, shows external similarities to *Modiolopsis alvae*, except for its more pronounced umbo and umbonal ridge. The general shape of *Ectenocaridomorpha* acuta is subtrigonal, while *Modiolopsis alvae* is typically elongate.

Ecology and habitat. – *Modiolopsis alvae* is principally streamlined, i.e. elongate and compressed, with an incouspicuous umbo that hardly reaches above the dorsal margin, and a smooth shell surface. Accordingly, based on external features alone, this species seems to have been well adapted for burrowing in soft substrates. In the diagram of Fig. 23, it falls within the region of rapid burrowing (Fig. 23:16).

However, its shell form, except for the non-sinuated ventral margin, also resembles that of modern endo-byssate taxa, e.g., *Modiolus americanus* Leach. The general pattern of muscular impressions of *Modiolopsis alvae* is reminiscent of that of Recent mytilids as well as that of several other Palaeozoic modioliform modiomorphids.

Modiolopsis alvae has a reduced anterior end, but the reduction is not as pronounced as in modern epibyssate taxa, such as *Mytilus*. The maximum convexity of the valves of *Modiolopsis alvae* is at about mid-length and mid-height of the shell, i.e. it lacks the typical triangular cross section with a maximum convexity below mid-height seen in epibyssate species.

Taking into account also the muscular pattern, *Modiolopsis alvae* shows characteristics typical of both rapid-burrowing species as well as of non-burrowing, endo-byssate species. I suggest that *M. alvae* was a weakly byssate. semi-infaunal burrowing species (Fig. 25).

The specimens from the Halla and Hemse Beds were found in extremely argillaceaous limestone, suggesting a soft original substrate. In such an environment a functional byssus is advantageous to keep a long shell in a safe life position (see discussion of *Modiodonta gothlandica*).

The Burgsvik specimens are smaller and shorter than the above mentioned ones and lived in a more unstable environment, as indicated by the oolitic limstone in which they were found.

Occurrence. – Wenlockian (Homerian) Halla Beds at Gothemshammar 7, Klinteberg Beds at Bryggans fiskeläge; Ludlovian Hemse Beds at Östergarn; Whitcliffian Burgsvik Beds, Gotland.

Summary

The Silurian Nuculoida of Gotland and Modiomorphidae of Sweden have been studied in terms of taxonomy, classification, functional morphology, and stratigraphical distribution. All available specimens of modiomorphids and nuculoids have been investigated, including several hundred specimens borrowed from the Swedish Museum of Natural History, Stockholm, and the Geological Survey of Sweden, Uppsala. Additional material from the Department of Historical Geology and Palaeontology, Lund is also included in this study. Comparisons have been made with actual North American material.

Because of the large quantity of nuculoid specimens and in most cases also excellent state of preservation, it has been possible to reconstruct soft-part anatomy and functional morphology of most species. Life habit and ecology of these bivalves have been suggested accordingly.

Three different nuculoid faunas have been studied with regard to life habit, including the reconstruction of feeding levels, or tiers, of the individual species.

From the silicified fauna in the Halla Beds at Möllbos, counting more than 3500 specimens, ontogenetic growth series for some of the nuculoid species have been constructed. Rare phenomena, such as traces of parasites and of predators, pearl building etc. have also been recorded. Tiers of this bivalve community have been proposed.

Twenty-one different species, assigned to 10 genera, are distributed among the different nuculoid families as follows: *Nuculodonta gotlandica*, *Ledopsis burgsvikensis*, *Similodonta djupvikensis*, and *Praenucula faba* in Praenuculidae; *Nuculoidea lens*, *Nuculoidea pinguis*, *Nuculoidea burgsvikensis*, *Nuculoidea* sp. A, ?*Nuculoidea ecaudata*, and ?*Nuculoidea* sp. in Nuculidae; *Caesariella lindensis*, *Ekstadia tricarinata*, *Ekstadia kellyi*, *Palaeostaba baltica*, *Nuculites solida*, *Nuculites* sp. A, and *Nuculites* sp. B in Malletiidae; '*Nuculana*' *oolitica* and '*Nuculana*' sp. A in Nuculanidae. *Tancrediopsis gotlandica* and *Tancrediopsis solituda* are questionably referred to Malletiidae (cf. Pojeta 1988, p. 211).

The nuculoid bivalves studied ranges from at least the Wenlockian Visby Beds to the Sundre beds of Gotland. Nuculoids have been found in all units on Gotland except for the Tofta Beds, the Klinteberg beds, the Eke Beds, and the Sundre Beds.

The different nuculoid species occur in lithologies laid down under various environmental conditions, from muds, fine-grained argillaceous limestones to coarse sandstones and oolitic limestones.

The Silurian modiomorphids of Sweden are equivalved to subequivalved bivalves of elongate subovate shell shape. They have a reduced anterior lobe and an opisthodetic elongate ligament, which resembles that of modern mytilids. They lack resilial ridges but instead they have grooves for the reception of the ligament.

They exhibit typically small beaks and umbones, and most species have smooth, compressed, stream-lined shells well fit for efficient burrowing. The valves have either a hinge plate without teeth or with one or several cardinal teeth, or are provided with a cardinal tooth without a hinge plate. Posterior teeth of the hinge are lacking.

The closest living relatives of the Modiomorphidae, the Mytilidae, have only a small, rudimentary foot and lack normal pedal burrowing mechanism. Accordingly, they are not effective burrowers (Stanley 1972, p. 171). The modiomorphids under study, however, exhibit a muscular morphology pattern intermediate between highly effective burrowers, like nuculoids, and Recent mytilids. It is therefore suggested that the Silurian modiomorphids examined retained the ability of effective burrowing, and several of them also had a byssus for maintaining a stable life position.

Four different modes of life have been suggested for the modiomorphids studied, viz. epifaunally epibyssate, semi-infaunally endobyssate, semi-infaunally non-byssate, and infaunally non-byssate (with fused mantle margins and presence of siphons).

Some 25 species of modiomorphids are described, of which 22 are new. These belong to seven genera, two of which are new. Two new subfamilies are suggested. The taxa are: Subfamilies Modiomorphinae and Modiolopsinae. *Modiodonta gothlandica* Liljedahl, 1989, *Colpomya hugini* n.sp., *C. munini* n.sp., ?*C. heimeri* n.sp., ?*C. vaki* n.sp., ?*C. ranae* n.sp., ?*C. audae* n.sp., ?*C. balderi* n.sp., ?*C. lokei* n.sp., ?*C. friggi* n.sp., ?*C.* sp. 1, *Aleodonta burei* n.gen and sp., *Mimerodonta atlei* n.gen. and sp., *M. njordi* n.gen. and sp., *Radiatodonta* sp. 1, *R.* sp. 2, *Goniophora onyx* Liljedahl, 1984, *G. bragei* n.sp., *G. brimeri* n.sp., *G. tyri* n.sp., *G. alei* n.sp., *G. acuta* Lindström, 1880, '*G.*' *gymeri* n.sp., '*G.*' sp. 1, and *Modiolopsis alvae* n.gen. and sp.

The modiomorphids are generally rare constituents of the Silurian faunal associations in Sweden. However, when present they may form a substantial part of the bivalve fauna, for instance, that in the Wenlockian Mulde marl at Djupvik, Gotland. They have been found in sediments indicating a wide range of environmental conditions. Fine-grained limestones and siltstones, indicating moderate to low wave and current action, contain the most prolific faunas (low diversity but high numbers of each taxon) while coarse-grained sediments such as oolitic limestones and sandstones have faunas of several species but with only small populations of each. Reef environments were less favourable since most species were semi-infaunal soft bottom dwellers.

The bivalves studied range from at least the Wenlockian Visby Beds of Gotland to the Pridolian Öved Sandstone in Scania. Modiomorphids have been found in all units on Gotland except for the Högklint Beds, the Tofta Beds, the Eke Beds, the Hamra Beds and the Sundre Beds. They occur only in the uppermost Silurian units in Scania, viz. the Colonus Shale, the Klinta Formation and the Öved Sandstone.

Four of the seven genera studied (*Aleodonta*, *Colpomya*, *Mimerodonta*, and *Modiodonta*) seem to have been endemic to Baltoscandia in Silurian time. However, because coeval British bivalve faunas have been described on external, and thus systematically unsatisfactory, characters only, it may well be that at least some of the endemic Gotland taxa had a wider geographic distribution than is now apparent.

Acknowledgments. – Sincere thanks go to Anita Löfgren, Stig M. Bergström, Robert C. Frey, John Pojeta Jr., Jiří Kříž, Peter Bengtson, and Mike Bassett for constructive criticism, to Euan Clarkson for linguistic help, to Sven Stridsberg for help with developing of films, Karen Taylor for drawing assistance, Leif Andersson for graphic help, and Adrian Rushton for providing information on the lectotype of *Goniophora cymbaeformis*, and Professor Valdar Jaanusson, Stockholm for the loan of material. The field work and travels to museums in Sweden and North America was made possible through grants from the Swedish Natural Science Research Council. Part of the study was carried out at the Ohio State University, and a Fulbright Commission award is greatly acknowledged for partly financing my stay there.

References

Adams, H. & Adams, A. 1858: *The Genera of Recent Mollusca.* 136 pp. London.

Allen, J.A. 1953: Observations on the epifauna of the deep-water muds of the Clyde Sea area, with special reference to *Clamys septemradiata* (Muller). *Journal of Animal Ecology 22*, 240–260.

Allen, J.A. 1954: A comparative study of the British species of *Nucula* and *Nuculana. Journal of Marine Biology Associations of the U.K. 33*, 457–472.

Angelin, N.P. & Lindström, G. 1880: *Fragmenta Silurica e Dono Caroli Henrici Wegelin.* 39 pp. Stockholm.

Babin, C. 1966: *Mollusques Bivalves et Céphalopods du Paléozoique Armoricain.* 470 pp. Brest.

Bailey, J.B. 1983: Middle Devonian bivalvia from the Solsville member (Marcellus Formation), Central New York State. *American Museum of Natural History, Bulletin 174*, 194–325.

Bailey, J.B. 1986: Systematics, hinge, and internal morphology of the Devonian bivalve *Nuculoidea corbuliformis* (Hall and Whitfield). *Journal of Paleontology 60*, 1177–1185.

[Bambach, R.K. 1969: Bivalvia of the Siluro-Devonian Arisaig Group, Nova Scotia. Unpublished thesis, Yale University, New Haven, Conn.]

Barrande, J. 1881: *Système Silurien du Centre de la Bohème, Vol. 6.* 342 pp. Prague.

Barrois, C.E. 1891: Mémoire sur la faune du grès armoricaine. *Annales de la Societé Géologique du Nord Tome 19*, 134–237.

Beushausen, H.E.L. 1884: Beiträge zur Kenntnis des Oberharzer Spiriferensandsteins und seiner Fauna. *Geologische Specialkarte Preussen und den Thuringen Staaten, Abhandlungen 61.* 133 pp.

Beushausen, H.E.L. 1895: Die Lammellibranchiaten der Rheinischen Devon mit Ausschluss der Aviculiden. *Königliche Preussische Geologische Landesanstalt, Abhandlungen Neue Folge 17.* 514 pp.

Billings, E. 1874: On some new genera and species of Paleozoic Mollusca. *Canadian Naturalist, New Series. 7*, 301–302.

Bowen, Z.P., Rhoads, D.C. & McAlester, A.L. 1974: Marine benthic communities in the Upper Devonian of New York. *Lethaia 7*, 93–120.

Bradshaw, J.D. & Bradshaw, M.A.1971: Functional morphology of some fossil palaeotaxodont bivalve hinges as a guide to orientation. *Palaeontology 14:2*, 242–249.

Bradshaw, M.A.1974: Morphology and mode of life of the bivalves *Nuculoidea vespa* n.sp. and *Nuculoidea umbra* n.sp. from the Devonian of New Zealand. *New Zealand Journal of Geology and Geophysics 17*, 447–464.

Bradshaw, M.A. 1978: Position of soft parts in fossil palaeotaxodont bivalves as suggested by features of the shell. *Alcheringa 2*, 203–215.

Bradshaw, M.A. 1979: Functional morhphology of some fossil palaeotaxodont bivalve hinges as a guide to orientation. *Palaeontology 14:2*, 242–249.

Bronn, H.G. 1832: *Ergebnisse Meiner Naturhistorisch-Ökonomischen Reisen. Pt. 2.* 686 pp. Heidelberg & Leipzig.

Chapman, F.A. 1908: Monograph of the Silurian bivalved mollusca of Victoria, in the collection of the National Museum, Melbourne. *Memoirs of the National Museum of Melbourne,2*, 5–62.

Conrad, T.A. 1841: On the Palaeontology of the State of New York. *New York State Geological Survey 5 annual report*, 116–118.

Cossman, M. 1897: Catalogue illustré des coquilles fossiles des environs de Paris. *Societé Royal Belgique, Annales 22:2.* 214 pp.

Cox, L.R. 1960: Thoughts on the classification of the Bivalvia. *Malacological Society of London, Proceedings, 34:2*, 60–88.

Cox, L.R. 1969: General features of the bivalvia. *Paleontology, Part N, Mollusca 6, Bivalvia 1*, N2–N121. Geological Society of America, Boulder, Colorado, and University of Kansas Press, Lawrence, Kansas.

Creer, K.M. 1973: A discussion of the arrangement of palaeomagnetic poles on the map of Pangaea for Epochs in the Phanerozoic. *In* Tarling, D.H. & Runcorn, S.K. (eds.): *Implications of Continental Drift to the Earth Sciences 1*, 47–46. Academic Press, London.

Dahmer, G. 1921: Studien über die Fauna des Oberharzer Kahlenbergsandsteins, II. *Preussische Geologische Landesanstalt, Berlin, 1919, 40, II, Heft 2*, 161–306.

Dahmer, G. 1936: Die Fauna der Obersten Siegener Schichten von Unkelmühle bei Eitorf a. d. Sieg. *Preussische Geologische Landesanstalt, Abhandlungen Neue Folge 168.* 36 pp.

Dall, W.H. 1889: On the hinge of pelecypods and its development, with an attempt toward a better subdivision of the group. *American Journal of Science 3*, 445–462.

Dall, W.H. 1895: Contributions to the Tertiary fauna of Florida and a classification of the pelecypods. *Wagner Free Institute of Science, Transactions, 3*, 483–947.

Dall, W.H. 1913: Pelecypoda. *In* von Zittel, K.A.: *Textbook of Palaeontology* (Eastman, C.R. transl. & ed.) Rev. Edit., Vol. 1, 421–507. McMillan, London.

Dell, R.K. 1987: Mollusca of the family Mytilidae (Bivalvia) associated with organic remains from deep water off New Zealand, with revisions of the genera *Adipicola* Dautzenberg, 1927 and *Idasola* Iredale, 1915. *National Museum of New Zealand, Records, 3*, 17–36.

Douvillé, H. 1913: Classification des Lamellibranchs. *Societé Géologique de France, Bulletin 12:4*, 419–467.

Drevermannn, F. 1902: Die Fauna der Untercoblenzschichten von Oberstadtfeld bei Daun in der Eifel. *Palaeontographica 49*, 78–94.

Drew, G.A. 1900: Locomotion in *Solenomya* and its relatives. *Anatomische Anzeige 17*, 15, 257–266.

Drew, G.A. 1901: The life history of *Nucula delphinodonta* (Mighels). *Quarternary Journal of Microscopical Sciences, 44*, 313–391.

Driscoll, E.G. 1964: Accessory muscle scars, an aid to protobranch orientation. *Journal of Paleontology 38*, 61–66.

Eichwald, E. 1860: *Lethaea Rossica ou Paléontologie de la Russie.* 1657 pp. Schweizerbart, Stuttgart.

Eriksson C.-O. & Laufeld, S. 1978: Philip structures in the submarine Silurian of northwest Gotland. *Sveriges Geologiska Undersökning C, 736.* 30 pp.

Férussac, A.E. de 1822: *Tableaux Systématiques des Animaux Mollusques.* 11 pp. Paris, London.

Fischer, P.H. 1886: Manuel de conchyliologie et de paléontologie conchyliologique. *Histoire Naturélle de Mollusque Vivants et Fossiles 10*, 897–1008.

Goldfuss, A. 1834–40: *Petrefacta Germaniae II, Abbildungen und Beschreibungen der Petrefacten Deutschlands und Angränzende Ländern.* 312 pp. Düsseldorf.

Gray, J.E. 1824: Shells. *In* W.E. Parry: *A Supplement to the Appendix of Captain Parry's Voyage for the Discovery of a North-West Passage, in the Years 1819–20,* 240–246. London.

Gray, J., Laufeld, S. & Boucot, A.J. 1974: Silurian trilete spores and spore tetrad from Gotland: Their implication for land plant evolution. *Science 185*, 260–263.

Grönwall, K.A. 1897: *In* Moberg, J.C. & Grönwall, K.A. 1897. Om Fyledalens Gotlandicum. *Lunds Universitets Årsskrift 5:1*, 86 pp.

Hall, J. 1843: Survey of the Fourth Geological District. *Natural History of New York, Geology of New York, Part 4.* 683 pp.

Hall, J. 1847: Palaeontology. *Geological Survey of New York 1.* 338 pp. Albany.

Hall, J. 1856: On the genus *Tellinomya* and allied genera. *Canadian Naturalist and Geologist 1*, 390–395.

Hall, J. 1869: *Preliminary Notice of the Lamellibranchiate Shell of the Upper Helderberg, Hamilton and Chemung Groups, With Others from the Waverly Sandstones. Part 2.* 97 pp. N.Y. State Museum, Albany.

Hall, J. 1870: On the relation of the Oneonta sandstone and Montrose sandstone of Vanuwem with the Hamilton and Chemung groups. *American Naturalist 4*, 563–565, 639–640.

Hall, J. 1883: Lamellibranchiata, plates and explanations. *Natural History of New York, Paleontology 5, part 1.* 20 pp.

Hall, J. 1885: Lamellibranchiata II, descriptions and figures of the Dimyaria of the upper Helderberg, Hamilton, Portage and Chemung Groups. *New York Geological Survey, Paleontology 5:1*, 1–268.

Hall, J. & Whitfield, R.P. 1875: Descriptions of Silurian fossils. *Geological Survey of Ohio, Report 2. 2. Paleontology*, 65–161.

Harrington, H.J. 1938: Las faunas del Ordoviciano inferior del norte Argentino. *La Plata Universita Nacional Institute Muse Revue, New Ser. 1, Sec. Paleontologia, 4*, 290 pp.

Heath, H. 1937: The anatomy of some protobranch molluscs. *Musée Royal d'Histoire Natural Belgique, Memoires 2*, 10, 1–26.

Hede, J.E. 1921: Gottlands silurstratigrafi. *Sveriges Geologiska Undersökning C 305.* 100 pp.

Hede, J.E.1925: Berggrunden (Silursystemet). *In* Munthe, H., Hede, J.E. & von Post, L. 1925: Beskrivning till kartbladet Ronehamn. *Sveriges Geologiska Undersökning Aa 156.* 34 pp.

Hede, J.E. 1927a: Berggrunden (Silursystemet). *In* Munthe, H., Hede, J.E. & Lundqvist, G. 1927: Beskrivning kartbladet Klintehamn. *Sveriges Geologiska Undersökning Aa 160.* 37 pp.

Hede, J.E. 1927b: Berggrunden (Silursystemet). *In* Munthe, H., Hede, J.E. & von Post, L. 1927: Beskrivning till kartbladet Hemse. *Sveriges Geologiska Undersökning Aa 164.* 36 pp.

Hede, J.E. 1940: Berggrunden. *In* Lundqvist, G., Hede, J.E. & Sundius, N. 1940. Beskrivning till kartbladet Visby och Lummelunda. *Sveriges Geologiska Undersökning Aa 183*, 9–68.

Hede, J.E. 1960: The Silurian of Gotland. *In* Regnéll, G. & Hede, J.E. 1960. The Lower Palaeozoic of Scania. The Silurian of Gotland. *International Geological Congress XXI Sess. Norden 1960 Guide to excursions Nos. A 22 and C 17*, 44–89. Also in *Publications of the Institute of Mineralogy, Palaeontology and Quaternary University of Lund 91*, 44–89.

Hicks, H. 1873: On the Tremadoc rocks in the neighbourhood of St. David's, south Wales, and their fossil contents. *Quarterly Journal of the Geological Society of London 29*, 39–52.

Hind, W. 1910: The Lamellibranchs of the Silurian Rocks of Girvan. *Transactions of the Royal Society of Edinburgh 47, Pt. 3:18*, 479–548.

Hisinger, W. 1827: Gottland, geognostiskt beskrifvit. *Kungliga Vetenskaps-Akademiens Handlingar för år 1826*, 311–337. Stockholm.

Hisinger, W. 1828: *Anteckningar i Physik och Geognosie, Fjärde häftet.* 258 pp. Stockholm.

Hisinger, W. 1831a: *Anteckningar i Physik och Geognosie under resor uti Sverige och Norrige. Femte häftet.* 174 pp. Stockholm.

Hisinger, W. 1831b [Published anonymously]: *Esquisse d'un Tableau des Pétrifications de la Suède. Nouvelle Edition.* 43 pp. Stockholm.

Hisinger, W. 1837: *Lethaea Svecica seu Petrificata Sveciae, Iconibus et Characteribus Illustrata.* 124 pp. Stockholm.

Hisinger, W., 1840: *Anteckningar i Physik och Geognosie under resor uti Sverige och Norrige. Sjunde häftet.*147 pp. Stockholm.

Hisinger, W. 1841: *Förteckning Öfver en Geognostisk och Petrefactologisk Samling från Sverige och Norrige.* 69 pp. Stockholm. (Reprinted 1842.)

Iredale, T. 1939: Great Barrier Reef expedition 1928–1929. *British Museum (Natural History), Scientific Reports, 5:6 (Mollusca, Pt. 1)*, 209–425.

Isberg, O. 1934: *Studien über Lamellibranchiaten des Leptaenakalkes in Dalarna.* 493 pp. Håkan Ohlssons, Lund.

Jackson, R.T. 1890: Phylogeny of the Pelecypoda. *Boston Society of Natural History Memoires 4*, 277–400.

[Jaanusson, V. 1986: Locality designations in old collections from the Silurian of Gotland. Department of Palaeozoology, Swedish Museum of Natural History, Stockholm. 19 pp. Stencil, unpublished.]

Jeppsson, L. 1974: Aspects of Late Silurian conodonts. *Fossils and Strata 6.* 54 pp.

Jeppsson, L. 1983: Silurian conodont faunas from Gotland. *Fossils and Strata 15.* 121–144.

Jeppsson, L., Fredholm, D. & Mattiasson, B. 1985: Acetic acid and phosphatic fossils – a warning. *Journal of Paleontology 59*, 952–956.

Jeppsson, L. & Laufeld, S. 1986: The late Silurian Öved–Ramsåsa group in Skåne, south Sweden. *Sveriges Geologiska Undersökning Ca 58.* 45 pp.

Kauffman, E.G. 1969: Form, function, and evolution. *In* Moore, R.C. (ed.): *Treatise on Invertebrate Paleontology, Part N, Mollusca 6, Bivalvia 1,* N129–N205. Geological Society of America, Boulder, Colorado, and University of Kansas Press, Lawrence, Kansas.

Kjerulf, T. 1865: *Veiviser ved Geologiske Excursioner i Christiania Omegn.* 43 pp. Brögger & Christie's.

Korobkov, I.A. 1954: *Spravochnik i Metodicheskoe Rukovodstvo po Tretichnym Mollyuskam Plastinchatozhabernye [Handbook on and Systematic Guide to the Teriary Mollusca, Lamellibranchiata.]* 444 pp. Gosud. Nauchno-tech. Issledov. Nefti. Gorno-toplivnoj lit-ry. Leningradskoe Otdelenie.

Kříž, J. & J. Pojeta 1974: Barrande's colonies concept and a comparison of his stratigraphy with the modern stratigraphy of the Middle Bohemian Lower Palaeozoic rocks (Barrandian) of Czechoslovakia. *Journal of Paleontology 48:3*, 489–494.

Larsson, K. 1979: Silurian tentaculitids from Gotland and Scania. *Fossils and Strata 11.* 180 pp.

Laufeld, S. 1974: Silurian Chitinozoa from Gotland. *Fossils and Strata 5.* 130 pp.

Laufeld, S., Bergström, J. & Warren, P. 1975: The boundary between the Silurian Cyrtograptus and Colonus Shales in Skåne, southern Sweden. *Geologiska Föreningens i Stockholm Förhandlingar 97*, 207–222.

Laufeld, S. & Jeppsson, L. 1976: Silicification and bentonites in the Silurian of Gotland. *Geologiska Föreningens i Stockholm Förhandlingar 98*, 31–44.

[LePennec, M. 1978: Génèse de la coquille larvaire et postlarvaire chez divers bivalves marins. Université de Bretagne, Brest. Unpublished dissertation, text and atlas. 229 pp.]

Liljedahl, L. 1983: Two silicified Silurian bivalves from Gotland. *Sveriges Geologiska Undersökning C 799*. 55 pp.

Liljedahl, L.1984a: Silurian silicified bivalves from Gotland. *Sveriges Geologiska Undersökning C 804*. 82 pp.

Liljedahl, L. 1984b: *Janeia silurica*, a link between nuculoids and solemyoids (Bivalvia). *Palaeontology 27*, 693–698.

Liljedahl, L. 1984c: Silicified Silurian bivalves from Gotland. *Lund Publications in Geology 24*. 28 pp.

Liljedahl, L. 1985: Ecological aspects of a silicified bivalve fauna from the Silurian of Gotland. *Lethaia 18*, 53–66.

Liljedahl, L. 1986: Endolithic micro-organisms and silicification of a bivalve fauna from the Silurian of Gotland. *Lethaia 19*, 267–278.

Liljedahl, L. 1989a: *Fylgia baltica* gen. et sp.nov. (Bivalvia) from the Silurian of Gotland. *Geologiska Föreningens i Stockholm Förhandlingar 111*, 339–345.

Liljedahl, L. 1989b: Identity of the bivalve *Modiodonta gothlandica* (Hisinger) from the Silurian of Gotland. *Geologiska Föreningens i Stockholm Förhandlingar 111*, 313–318.

Liljedahl, L. 1989c: Two micromorphic bivalves from the Silurian of Gotland. *Paläontologische Zeitschrift 63*, 229–240.

Liljedahl, L. 1991: Contrasting feeding strategies in bivalves from the Silurian of Gotland. *Palaeontology 34*, 17.

Liljedahl, L. 1992a: *Silurozodus*, new genus, oldest known member of the Trigonioida (Bivalvia, Mollusca). *Paläontologische Zeitschrift 66*, 51–65.

Liljedahl, L. 1992b: *Yonginella*, a new bivalve (Mollusca) from the Silurian of Gotland. *Journal of Paleontology 66*, 211–214.

Liljedahl, L. 1992c: The Silurian bivalve *Ilionia prisca*, oldest known deep-burrowing suspension feeding bivalve. *Journal of Paleontology 66*, 206–210.

Lindström, G. 1880: *In* Angelin, N.P. & Lindström, G. 1880: *Fragmenta Silurica e Dono Caroli Henrici Wegelin*. 39 pp. Stockholm.

Lindström, G. 1882: Anteckningar om Silurlagern på Carlsöarne. *Öfversikt af Kungliga Vetenskaps-Akademiens Förhandlingar 3*, 5–30.

Lindström, G. 1885: List of the fossils of the Upper Silurian Formation of Gotland. 20 pp. Norstedt, Stockholm.

Lindström, G. 1888: *List of the Fossil Faunas of Sweden. I. Cambrian and Lower Silurian*. 12 pp. Stockholm.

Linnaeus, C. 1758: *Systema Naturae per Regna Tria Naturae. Edit. Decima, 1, ii.* 824 pp. Holmiae.

McAlester, A.L. 1962: Upper Devonian pelecypods of the New York Chemung Stage. *Peabody Museum of Natural History, Yale University Bulletin 16*. 88 pp.

McAlester, A.L. 1963: Revison of the type species of the Ordovician nuculoid pelecypod genus *Tancrediopsis*. *Postilla 74*. 19 pp. Yale Peabody Museum of Natural History, New Haven.

McAlester, A.L. 1968: Type species of Palaeozoic nuculoid bivalve genera. *Geological Society of America Memoir 105*, 143 pp.

McAlester, A.L. 1969: Superfamily Ctenodontacea. Diagnosis Superfamily Nuculacea, and Diagnosis Superfamily Nuculanacea. *In* Moore, R.C. (ed.): *Treatise on Invertebrate Paleontology, Part N, Mollusca 6, Bivalvia 1*, N227, N229, N231. Geological Society of America, Boulder, Colorado, and University of Kansas Press, Lawrence, Kansas.

McCoy, F. 1855: Descriptions of the British Palaeozoic fossils. *In* Sedgwick, A. 1855: *A Synopsis of the Classification of the British Palaeozoic Rocks*. 661 pp. London and Cambridge.

McCoy, F. 1862: *A Synopsis of the Silurian Fossils of Ireland*. 72 pp. Dublin.

McLearn, F.H. 1918: The Silurian Arisaig Series of Arisaig, Nova Scotia. *American Journal of Science 45*, 126–140.

McLearn, F.H. 1924: Palaeontology of the Silurian rocks of Arisaig, Nova Scotia. *Geological Survey of Cananada Memoir 137*. 180 pp.

Mallieux, E. 1937: Les lamellibranchs du Dévonien Inférieur de l'Ardenne. *Musée Royal Histoire Natural Belgique, Memoires 81*. 273 pp.

Manten, A.A. 1971: Silurian reefs of Gotland. *Developments in Sedimentology 13*. 539 pp. Elsevier, Amsterdam.

Martinsson, A. 1962: Ostracodes of the family Beyrichiidae from the Silurian of Gotland. *Bulletin of the Geological Institute of Uppsala 41*. 369 pp.

Martinsson, A. 1967: The succession and correlation of ostracode faunas in the Silurian of Gotland. *Geologiska Föreningens i Stockholm Förhandlingar 89*, 350–386.

Miller, S.A. 1877: *The American Palaeozoic Fossils, a Catalogue of the Genera and Species*. 253 pp. Cincinnati.

Miller, S.A. 1889: *North American Geology and Palaeontology*. 664 pp. Cincinnati.

Moberg, J.C. & Grönwall, K.A. 1909: Om Fyledalens Gotlandicum. *Meddelanden från Lunds Geologiska Fältklubb B 3*. 86 pp.

Murchison, R.I. 1839: *The Silurian System – 2*. 628–629. John Murray, London.

Murchison, R.I. 1855: *Siluria. A History of the Oldest Rocks in the British Isles and Other Countries*. 566 pp. London.

Newell, N.D. 1942: Late Paleozoic pelecypods: Mytilacea. *Kansas Geological Survey Report 10:2*. 110 pp.

Newell, N.D. 1957: Notes on certain primitive heterodont pelecypods. *American Museum Novitates 1857*. 14 pp.

Newell, N.D. 1965: Classification of the Bivalvia. *American Museum Novitates 2206*. 25 pp.

Newell, N.D. 1969: Systematic descriptions. *In* Moore, R.C. (ed.): *Treatise on Invertebrate Paleontology, Part N, Mollusca 6, Bivalvia 1*, N213–N241. Geological Society of America, Boulder, Colorado, and University of Kansas Press, Lawrence, Kansas.

d'Orbigny, A. 1850: *Prodrome de Paléontologie Stratigraphique Universelle 1*. 394 pp. Paris.

Northrope, S.A. 1939: Paleontology and stratigraphy of the Silurian rocks of Port Daniel–Black Cape region, Gaspé. *Geological Society of America Special Papers 21*. 302 pp.

Ockelmann, K.W. 1983: Descriptions of mytilid species and definition of the Dacrydiinae n.subfam. (Mytilacea–Bivalvia). *Ophelia 22*, 81–123.

Owen, G. 1958: Shell form, pallial attachment and the ligament in the Bivalvia. *Zoological Society of London, Proceedings 131*, 637–648.

Owen, G. 1959: The ligament and digestive system in the taxodont bivalves. *Malacological Society of London Proceedings 33*, 215–223.

Owen, G., Trueman, E.R. & Yonge, C.M. 1953: The ligament in the Lamellibranchia. *Nature 171*, 73–75.

Pelseneer, P. 1891: Contribution a l'étude des Lamellibranches. *Archives de Biologie 11*, 147–312.

Pfab, L. 1934: Revision der Taxodonta des Böhmischen Silurs. *Palaeontographica A*, 195–253.

Philip, G.M. 1962: The paleontology and stratigraphy of the Siluro-Devonian sediments of the Tyers area, Gippsland, Victoria. *Royal Society of Victoria, Proceedings, University Series 75*, 123–246.

Phillips, J. & Salter, J.W. 1848: Palaeontological appendix to Professor John Phillips' memoir on the Malvern Hills, compared with the Palaeozoic districts of Abberley, Woolhope, May Hill, Tortworth, and Usk. *Geological Survey of Great Britain, Memoirs 2:1*, 359–369.

Pojeta, J. Jr. 1966: North American Ambonychiidae (Pelecypoda). *Palaeontographica Americana 5:36*, 129–241.

Pojeta, J. Jr. 1971: Review of Ordovician pelecypods. *U.S. Geological Survey Professional Paper 695*, 1–46.

Pojeta, J. Jr. 1978: The origin and early taxonomic diversification of pelecypods. *Philosophical Transactions Royal Society of London B 284*, 225–246.

Pojeta, J. Jr. 1979: Pelecypods. *In* Jaanusson, V., Laufeld, S. & Skoglund, R. (eds.): Lower Wenlock faunal and floral dynamics – Vattenfallet section, Gotland. *Sveriges Geologiska Undersökning C 762*, 109–112.

Pojeta, J. Jr. 1985: Early history of diasome mollusks. *In* Bottjer, D.J., Hickman, C.S. & Ward, P.D. (eds.): Mollusks, Notes for a short course. *University of Tennesee Deptartment of Geological Sciences, Studies in Geology 13*, 102–130.

Pojeta, J. Jr. 1988: The origin and Paleozoic diversification of solemyoid pelecypods. *New Mexico Bureiau of Mines & Mineral Resources Memoir 44*, 201–271.

Pojeta, J. Jr. & Gilbert-Tomlinson J. 1977: Australian Ordovician pelecypod molluscs. *Department of Natural Resources Bureau of Mineral Resources, Geology & Geophysics, Bulletin 174*. 64 pp.

Pojeta, J. Jr. & Runnegar, B. 1985: The early evolution of Diasome Molluscs. *The Mollusca, 10, Evolution*, 295–336.

Pojeta, J. Jr., Zhang, R. & Yang, Z. 1986: Systematic paleontology of the Devonian pelecypods of Guangxi and Michigan. *In* Pojeta, J. Jr. (ed): Devonian rocks and Lower and Middle Devonian pelecypods of Guangxi, China, and the Traverse Group of Michigan. *U.S. Geological Survey Professional paper 1394*, 57–108.

Portlock, A. 1843: *Report on the Geology of Londonderry, and Parts of Tyrione and Fermanagh. i–xxxi, A-I*. 784 pp. Dublin and London.

Prantl, F. & Růžička, B. 1954: *Straba* nov.gen., a new Devonian pelecypod from Bohemia. *Acta Musei Nationalis Prague 10:3*, 29 pp.

Quenstedt, W. 1929: Mollusken aus der Redbay- und Greyhook-Schichten Spitzbergens. *Resultater av de norske statsunderstöttede Spitzbergen-expeditioner (Skrifter om Svalbard og Ishavet) 1:11*, 36–41

Rafinesque, C.S. 1815: *Analyse de la Nature*. 225 pp. Palermo.

Reed, F.R.C. 1927: Palaeontological notes on the Silurian inlier of Woolhope. *Quarterly Journal of the Geological Society of London 83*, 531–549.

Reed, F.R.C. 1931: New lamellibranchs from the Silurian of the Ludlow district. *Annales and Magazine of Natural History 8*, 290–303.

Rhoads, D.C. 1970: Mass properties, stability, and ecology of marine muds related to burrowing activity. *In* Crimes, T.P. & Harper, J.C. (eds.): *Trace Fossils*, 391–406. Seele House Press, Liverpool.

Rhoads, D.C. & Young, D.K. 1970: The influence of deposit-feeding organisms on sediment stability and community trophic structure. *Journal of Marine Research 28*:150–178.

Roemer, A. 1843: *Die Versteinigungen des Harzgebirges*. 40 pp. Hannover.

Runnegar, B. & Bentley, C. 1983: Anatomy, ecology, and affinities of the Australian Early Cambrian bivalve *Pojetaia runnegari* Jell. *Journal of Paleontology 47:1*, 73–92.

Salter, J.W. 1852: Note on the fossils above mentioned from the Ottowa River. *British Association of for the Advancement of Science, 21st Meeting Transactions, 1851 Report*, 63–65.

Sandberger, F. 1850: Versteinerungen des Schichtensystemes in Nassau. *Jahrbuch des Nassauischer Verein für Naturkunde 21*, 1–68.

Savazzi, E., 1984: Adaptive significance of shell torsion in mytilid bivalves. *Palaeontology 27*, 307–314.

Scotese, C.R. & McKerrow, W.S. 1990. Revised world maps and introduction. *In* McKerrow, W.S. & Scotese, C.R.: Palaeogeography and biogeography. *Geological Society of London Memoires 12*, 1–21.

Sharpe, D. 1853: Description of new species of Zoophyta and Mollusca. Appendix B to Ribeiro, C.: On the Carboniferous and Silurian Formations of the neighbourhood of Bussaco in Portugal. *Quarterly Journal of the Geological Society of London 9*. 185 pp.

Sherrard, K. 1959: Some Silurian lamellibranchs from New South Wales. *Proceedings of the Linnean Society of New South Wales 84*, 356–372.

Simpson, G.B. 1890: Descriptions of new species of fossils from the Clinton, Lower Helderberg, Chemung, and Waverly groups. *Transactions of the American Philosophical Society, New Series 16*, 435–460.

Soot-Ryen, T. 1955: A report on the family Mytilidae (Pelecypoda). *Allan Hancock Pacific Expeditions 20:1*. 175 pp.

Soot-Ryen, H. 1964: Nuculoid pelecypods from the Silurian of Gotland. *Arkiv för Mineralogi, Svenska Vetenskaps-Akademien, 3:28*, 489–519.

Sowerby, J. de C. 1839. *In* Murchison, R.I.: *The Silurian System 2*, 602–712. John Murray, London.

Stanley, S.M. 1970: Relation of shell form to life habits of the Bivalvia (Mollusca). *Geological Society of America Memoirs 125*, 125 pp.

Stanley, S.M. 1972: Functional morphology and evolution of byssally attached bivalve mollusks. *Journal of Paleontology 46:2*, 165–212.

Stevens, R.A. 1858: Descriptions of new Carboniferous fossils from the Appalachian, Illinois and Michigan Coal-fields. *American Journal of Science, 2nd Series, 25*, 258–265.

Termier, G. & Termier, H. 1971: Sur un bivalve protobranch ante-Arenigien trouvé au nord de Wardak (Afghanistan). *Geobios 4*, 143–150.

Thayer, 1974: Marine paleoecology in the Upper Devonian of New York. *Lethaia 7*, 121–155.

Thayer, C.W. 1975: Morphologic adaptions of benthic invertebrates to soft substrata. *Journal of Marine Research 33*, 177–189.

Thoral, M. 1935: *Contribution à l'Étude Paléontologique de l'Ordovicien Inférieur de la Montagne Noire et Révision Sommaire de la Fauna Cambrienne de la Montagne Noire*. Montpellier, 326 pp.

Trueman, E.R. 1950: Observations on the ligament of *Mytilus edulis*. *Quarterly Journal of Microscopical Sciences 91:3*, 225–235.

Trueman, E.R. 1969: General features of Bivalvia. *In* Moore, R.C. (ed.): *Treatise on Invertebrate Paleontology, Part N, Mollusca 6, Bivalvia 1*, N58–N64. Geological Society of America, Boulder, Colorado, and University of Kansas Press, Lawrence, Kansas.

Tunnicliff, S. 1982: A revision of late Ordovician bivalves from Pomeroy, Co. Tyrone, Ireland. *Palaeontology 25:1*, 43–88.

Twenhofel, W.H. 1927: Geology of Anticosti Island. *Geological Survey of Canada, Memoir 154*. 481 pp.

Ulrich, E.O. 1890: New Lamellibranchiata 1. *American Geologist 5:5*, 270–284.

Ulrich, E.O. 1892: New Lamellibranchiata 4, descriptions of one new genus and eight new species. *American Geologist 10*, 96–104.

Ulrich, E.O. 1893: New and little known Lamellibranchiata from the Lower Silurian rocks of Ohio and adjacent states. *Ohio Geological Survey Report 7:2*, 627–693.

Ulrich, E.O. 1894: The Lower Silurian Lamellibranchiata of Minnesota. From Vol. 3 of the Final Report. *Minnesota Geological and Natural History Survey*, 475–628.

Ulrich, E.O. 1897: The Lower Silurian Lamellibranchiata of Minnesota. *The Geological and Natural History Survey of Minnesota, Vol. 3:2 of the First Report*, 475–628.

Vokes, H.E. 1949: The hinge and marginal pectination of *Nuculoidea opima* (Hall), type of *Nuculoidea* Williams and Breger. *Journal of the Washington Academy of Science 39*, 361–363.

Vokes, H.E. 1980: *Genera of the Bivalvia. A Systematic and Bibliographic Catalogue*. 306 pp. Paleontological Research Institute, Ithaca, New York.

Williams, H.S. 1912: Some new Mollusca from the Silurian formations of Washington county, Maine. *Proceedings of the U.S. National Museum 42*, 381–401.

Williams, H.S. 1913: New species of Silurian fossils from the Edmunds and Pembroke formations of Washington County, Maine. *Proceedings of the U.S. National Museum 45*, 319–352.

Williams, H.S. & Breger, C.L. 1916: The fauna of the Chapman Sandstone of Maine including descriptions of some related species from the Moose River Sandstone. *U.S. Geological Survey Professional Paper 89*, 347 pp.

Yonge, C.M. 1941: The protobranchiate Mollusca: A functional interpretation of their structure and evolution. *Philosophical Transactions of the Royal Society of London B 230*:70–147.

Zhang, R.J. 1977: [Bivalvia. In Palaeontological Atlas of Central South China, Part 1], 73. Geological Press, Beijing. (In Chinese.)